SACRAMENTO PUBLIC LIBRARY
D0338331

SCATTER, ADAPT, AND REMEMBER

ALSO BY ANNALEE NEWITZ

She's Such a Geek!: Women Write About Science, Technology, and Other Nerdy Stuff (ed.)

Pretend We're Dead: Capitalist Monsters in American Pop Culture

The Bad Subjects Anthology (ed.)

White Trash: Race and Class in America (ed.)

ANNALEE NEWITZ

Scatter, Adapt, and Remember

How Humans Will Survive a Mass Extinction

DOUBLEDAY
New York London Toronto Sydney Auckland

www.doubleday.com

DOUBLEDAY and the portrayal of an anchor with a dolphin are registered trademarks of Random House, Inc.

Portions of Chapters 3, 6, and 16 are adapted from pieces originally published on io9.com.

Book design by Michael Collica
Jacket design by Emily Mahon
Jacket illustration by neilwebb.net

Library of Congress Cataloging-in-Publication Data
Newitz, Annalee, 1969–
Scatter, adapt, and remember : how humans will survive a mass extinction / Annalee Newitz.
pages cm
1. Survival. 2. Extinction (Biology) I. Title.
GF86.N485 2013
576.8'4–dc23 2012042409

ISBN 978-0-385-53591-5

MANUFACTURED IN THE UNITED STATES OF AMERICA

10 9 8 7 6 5 4 3 2 1

First Edition

To Charlie for the words
To Chris for the sounds
To Jesse for the stars

Utopian speculations . . . must come back into fashion.
They are a way of affirming faith in the possibility of solving
problems that seem at the moment insoluble. Today
even the survival of humanity is a utopian hope.

—Norman O. Brown, from *Life Against Death*

CONTENTS

SCATTER, ADAPT, AND REMEMBER

INTRODUCTION: *Are We All Going to Die?*

HUMANITY IS AT a crossroads. We have ample evidence that Earth is headed for disaster, and for the first time in history we have the ability to prevent that disaster from wiping us out. Whether the disaster is caused by humans or by nature, it is inevitable. But our doom is not. How can I say that with so much certainty? Because the world has been almost completely destroyed at least half a dozen times already in Earth's 4.5-billion-year history, and every single time there have been survivors. Earth has been shattered by asteroid impacts, choked by extreme greenhouse gases, locked up in ice, bombarded with cosmic radiation, and ripped open by megavolcanoes so enormous they are almost unimaginable. Each of these disasters caused mass extinctions, during which more than 75 percent of the species on Earth died out. And yet *every single time,* living creatures carried on, adapting to survive under the harshest of conditions.

My hope for the future of humanity is therefore not simply a warm feeling I have about how awesome we are. It is based on hard evidence gleaned from the history of survival on Earth. This book is about how life has survived mass extinctions so far. But it is also about the future, and what we need to do to make sure humans don't perish in the next one.

During the last million years of our evolution as a species, humans narrowly avoided extinction more than once. We lived through harsh conditions while another human group, the Neanderthals, did not. This isn't just because we are lucky. It's because as a species, we are extremely cunning when it comes to survival. If we want to survive for another million years, we should look to our history to find strategies that already worked. The title of this book, *Scatter, Adapt, and Remember,* is a distillation of these strategies. But it's also a call to implement them in the future, by actively taking on the project of human survival as a social and scientific challenge.

In the near term, we need to improve one of humanity's greatest inventions, the city, to make urban life healthier and more environmentally sustainable. Essentially, we need to adapt the metropolis to Earth's current ecosystems so that we can maintain our food supplies and a habitable climate. But even if you're not worried about climate change, Earth is still a dangerous place. At any time, we could be hit by an asteroid or a gamma-ray burst from space. That's why we need a long-term plan to get humanity off Earth. We need cities beyond the Blue Marble, oases on other worlds where we can scatter to survive even cosmic disasters.

But none of this will be possible if we don't remember human history, from our earliest ancestors' discovery of fire to our grandparents' development of space programs. Fundamentally, we are a species of builders and explorers. We've survived this long by taking control of our destiny. If we want to survive the next mass extinction, we can't forget how we got here. Now let's forge ahead into the future that we'll build for ourselves, our planet, and the humans who will exist a million years from now.

Evidence for the Next Mass Extinction

Over the past four years, bee colonies have undergone a disturbing transformation. As helpless beekeepers looked on, the machinelike efficiency of these communal insects devolved into inexplicable disorganization. Worker bees would fly away, never to return; adolescent bees wandered aimlessly in the hive; and the daily jobs in the colony were left undone until honey production stopped and eggs died of neglect. In reports to agriculture experts, beekeepers sometimes called the results "a dead hive without dead bodies." The problem became so widespread that scientists gave it a name—Colony Collapse Disorder—and according to the U.S. Department of Agriculture, the syndrome has claimed roughly 30 percent of bee colonies every winter since 2007. As biologists scramble to understand the causes, suggesting everything from fungal infections to parasites and pollution, farmers worry that the bee population will collapse into total extinction. If bees go extinct, their loss will trigger an extinction domino effect because crops from apples to broccoli rely on these insects for pollination.

At the same time, over a third of the world's amphibian species are threatened with extinction, too, leading many researchers to call this the era of amphibian crisis. But the crisis isn't just decimating bees and frogs. The Harvard evolutionary biologist and conservationist E. O. Wilson estimates that 27,000 species of all kinds go extinct per year.

Are we in the first act of a mass extinction that will end in the death of millions of plant and animal species across the planet, including us?

That's what proponents of the "sixth extinction" theory believe. As the term suggests, our planet has been through five mass extinctions before. The dinosaur extinction was the most recent but hardly the most deadly: 65 million years ago, dinosaurs were among the 76 percent of all species on Earth that were extinguished after a series of natural disasters. But 185 million years before that, there was a mass extinction so devastating that paleontologists have nicknamed it the Great Dying. At that time, 95 percent of all species on the planet were wiped out over a span of roughly 100,000 years—most likely from megavolcanoes that erupted for centuries in Siberia, slowly turning the atmosphere to poison. And three more mass extinctions, some dating back over 400 million years, were caused by ice ages, invasive species, and radiation bombardment from space.

The term "sixth extinction" was coined in the 1990s by the paleontologist Richard Leakey. At that time, he wrote a book about how this new mass extinction began 15,000 years ago, when the Americas teemed with mammoths, as well as giant elk and sloths. These turbo-vegetarians were hunted by equally large carnivores, including the saber-toothed cat, whose eight-inch fangs emerged from between the big cat's lips, curving to well beneath its chin. But shortly after humans' arrival on these continents, the megafauna populations collapsed. Leakey believes human habitat destruction was to blame for the extinctions thousands of years ago, just as it can be blamed today for the amphibian crisis. Leakey's rallying cry has resulted in sober scientific papers today, where respected biologists detail the evidence of a mass extinction in the making. *The New Yorker*'s environmental journalist Elizabeth Kolbert has tirelessly reported on scientific evidence gathered over the past two decades corroborating the idea that we might be living through the early days of a new mass extinction.

Though some mass extinctions happen quickly, most take hundreds of thousands of years. So how would we know whether one was happening right now? The simple answer is that we can't be sure. What we do know, however, is that mass extinctions have decimated our planet on a regular basis throughout its history. The Great Dying involved climate change similar to the one our planet is undergoing right now. Other extinctions may have been caused by radiation bombardment or stray asteroids, but as we'll see in the first section of this book, these disasters' most devastating effects involved environmental changes, too.

My point is that regardless of whether humans are responsible for the sixth mass extinction on Earth, it's going to happen. Assigning blame is less important than figuring out how to prepare for the inevitable and survive it. And when I say "survive it," I don't mean as humans alone on a world gone to hell. Survival must include the entire planet, and its myriad ecosystems, because those are what keep us fed and healthy.

There are many ways we can respond to the end of the world as we know it, but our first instincts are usually paralysis and depression. After all, what can you do about a comet hurtling towards us through space, unless you're Bruce Willis and his crack team of super-astronauts on a mission to blow that sucker up with a bunch of nukes? And what can you do to stop global environmental changes? This kind of "nothing can be done" response is completely understandable, but it rarely leads to pragmatic ideas about how to save ourselves. Instead, we are left imagining what the world will be like without us. We try to persuade ourselves that maybe things really will be better if humans just don't make it.

I'm not ready to give up like that, and I hope you aren't either. Let's assume that humans are just getting started on their long evolutionary trek through time. How do we switch gears into survival mode?

Survivalism vs. Survival

Many of us already have concrete ideas about how we'd survive a disaster. Survivalist groups build shelters stocked with food, preparing for everything from nuclear attack to super-storms. Most of us are survivalists in small ways, too, even if we don't build elaborate mountain hideaways. I live in San Francisco, where it's common for people to keep big jugs of water and food supplies in our homes just in case we're hit with a major earthquake. Our city government recommends that we all stash away enough supplies for a week, including fuel and water-purification tablets. Living here, I'm always aware of the possibility that my city might be in ruins tomorrow. It's such an ever-present danger that I've worked out a quake contingency plan with my family: If a large quake hits and we can't reach each other by phone, we're going to meet in the southwest corner of Dolores Park, an open area that's likely to be relatively safe and undamaged. We picked this location partly because over 100 years ago, people who survived San Francisco's last great quake met in Dolores Park, too.

One reason I decided to write this book is that I've spent so much time thinking about future disasters. I don't just mean the quake that's going to wreck my home. For most of my life I've been obsessed with stories about the end of the world. The whole thing probably started with the Godzilla movies I watched as a kid with my dad, but by the time I was an adult I'd consumed every story about the apocalypse I could get my hands on, from cheesy movies like *Hell Comes to Frogtown* to literary novels like Margaret Atwood's *Oryx and Crake*. When I was getting my Ph.D. in English, I wrote my dissertation on violent monster stories, exploring why people are drawn to the same tales of disaster over and over again. Eventually I left academia to become a science journalist, which didn't exactly curb my appetite for destruction. I produced stories about everything from computer hacking to pandemics. While I was at MIT doing a Knight Science Journalism fellowship, I was first exposed to the idea that planetwide mass extinction is a vital part of Earth's history, and an inevitable part of our future, too. Everything I had read in the fields of fiction and science led me to a single, dark conclusion. Humans are screwed, and so is our planet.

And so, a few years ago, I set out to write a book about how we are all doomed. I even printed out a brief outline of what I would research, then scribbled at the bottom: "Life is still nasty, brutish and short." With this idea in mind, I immersed myself in the scientific literature on mass extinction. But soon I discovered something I didn't expect–a single, bright narrative thread that ran through every story of death. That thread was survival. No matter how horrific things got, in geological and human history, life endured. I began to experience a kind of guarded optimism; perhaps billions of creatures would die in the coming mass extinction, but some would live. I reexamined my assumptions, and started to research what it would take for humans to be part of that bright narrative thread. I interviewed over a hundred people in fields from physics and geology to history and anthropology; I read about survival strategies in scientific journals, engineering manuals, and science-fiction novels; and I traveled all over the world to find evidence of humans' quest to survive in ancient cities and modern-day labs. I emerged from my research with the belief that humanity has a lot more than a fighting chance at making it for another million years.

Human beings may be experts at destroying life, including our own, but we are also tremendously talented at preserving it. For all the stories about human selfishness and bloodlust, there are just as many about people putting themselves in mortal danger to rescue strangers from burning houses or oppressive governments. Our urge to live spills over onto everything else around us: We don't want to live alone. During terrible disasters, we try to save as many other creatures as possible when we save ourselves. The urge to survive, not just as individuals but as a society and an ecosystem, is built into us as deeply as greed and cynicism are. Perhaps even more deeply, since the quest for survival is as old as life itself.

It's hard to convey in words what it's like to experience a change of heart based on gathering scientific evidence. I found hope in the historical accounts of human survival that Rebecca Solnit presents in *A Paradise Built in Hell: The Extraordinary Communities That Arise in Disaster*, and I found a scientific basis for that hope in Joan Roughgarden's *The Genial Gene: Deconstructing Darwinian Selfishness*. These thinkers and many more suggest we possess the cultural and evolutionary drive to help each other survive. Put another way, I gained a new appreciation for movies like *The Avengers*, where our heroes unite to save the world.

All survival strategies, however small, are signs that we harbor hope about the future. The problem is that most of our strategies, like my earthquake plan, are focused on personal survival. I'm only prepared to help myself and a few close companions make it through the coming disaster. Stashing away a week's worth of canned goods isn't a plan that scales well for an entire planet and all the human civilizations on it. Though it's not a bad idea to stock shelters with supplies for our families, we aren't going to survive a mass extinction that way. Our strategies need to be much bigger.

We have to move from survivalist tactics, aimed at protecting individual lives in a disaster, to survival strategies that could help our entire species make it through a mass extinction.

Learning from the Past

Though this shift in strategy sounds like a daunting task, we can take comfort in knowing that our early ancestors faced near-extinction too. In part one of this book, we'll plunge into geological deep time, and explore how life has endured through some of the most terrifying mass extinctions that have hit the planet over the past billion years. Then, in part two, we'll turn to the history of human evolution, and all its perils. Some anthropologists believe *Homo sapiens* struggled through a population bottleneck that brought our numbers down to thousands of individuals less than 100,000 years ago—possibly due to climate change, or simply from the hardships we faced as we migrated out of Africa. Regardless of what caused the population bottleneck, both the fossil record and genetic analysis suggest that humans were at one time rather sparse upon the Earth. To survive, we adopted strategies similar to other species that lived through centuries of poison skies and gigantic explosions. And one of those basic strategies was adaptability.

"Adaptability" is a term you hear a lot from people who study mass extinction. They talk about it with a weird, gallows-humor kind of optimism. This is evident when you meet Earth scientist Mike Benton, who has spent the past ten years studying the Great Dying and its survivors. In his line of work, Benton has sifted through the remains of some serious planetwide horrors. Two hundred and fifty million years ago, when the Great Dying happened, megavolcanoes fouled the atmosphere with carbon, and it's possible that an asteroid hit the planet, too. Despite Benton's intimate familiarity with mass death, he still maintains hope that our species will survive. He told me that "good survival characteristics for any animal" include being able to eat a lot of different things and live anywhere, just as humans can. Of course, he noted, that doesn't mean there won't be a lot of casualties. He continued:

> Evidence from mass extinctions of the past is that the initial killing is often quite random, and so nothing in particular can protect you, but then in the following grim times, when Earth conditions may still be ghastly, it's the adaptable forms that breed fast and live at high population size that have the best chance of fighting through.

We have a fighting chance because our population is large, plus we can adapt to new territories and eat a wide range of things. That's a good start, but what does it really mean to fight through? In part three of this book, we'll look at some specific examples of how humans and other creatures have used the three survival strategies of scattering, adapting, and remembering. We'll also explore how humans survive by planning for the future through storytelling. Fiction about tomorrow can provide a symbolic map that tells us where we want to go.

Stories of the Future

So where, exactly, do we want to go? With parts four and five, we'll launch ourselves into humanity's possible future. One of our biggest problems as a species today is that we have become so populous that our mass societies are no longer adaptive. Over half the population lives in cities, but cities can become death traps during disasters, and they are breeding grounds for pandemics. Worse, they are not sustainable; cities' energy and agricultural needs are outpacing availability, which limits their life spans and those of the people in them. Part four is about several ways we'll want to change cities over the next century to make them healthy, sustainable places that preserve human life as well as the life of the environment.

Often, a city-saving idea can start in a lab. Right now, in a cavernous warehouse on the Oregon State University campus, a group of researchers is designing the deadliest tsunami in history. In this cold, windy laboratory, they've got a massive water tank, about the size of an Olympic swimming pool, whose currents are controlled by a set of paddles bigger than doors. In the tank, wave after wave buffets a very detailed model city, washing away tiny wooden houses. Whirling in the water are special particles that can be tracked by hundreds of motion detectors, which help scientists understand tsunami behavior. At the tsunami lab, civil engineers destroy cities to figure out the best places for flood drains and high-ground emergency pathways in coastal cities.

Thousands of kilometers across the country, a revolutionary group of architects is working with biologists to create materials for "living cities" that are environmentally sustainable because they are literally part of the environment. These buildings might have walls made from semipermeable membranes that allow air in, along with a bit of rainwater for ceiling lights made from luminescent algae. Urbanites would grow fuel in home bioreactors, and tend air-purifying mold that flourishes around their windows. Unlike today's cities, these living cities will run on biofuels and solar energy. These are the kinds of metropolises where we and our ecoystems could thrive for millennia.

In part five, we'll look to the far future of humanity and think about our long-term plan to keep our species going for another million years. We know that when early humans were threatened with extinction they fanned out across Africa in search of new homes, eventually leaving the continent entirely. This urge to break away from home and wander has saved us before and could save us in the future. If we colonize other planets, then we will be imitating the survival strategy of our ancestors. Scattering to the stars echoes our journey out of Africa–and it could be our best hope for lasting through the eons.

Engineers at NASA are already preparing more robotic missions to the Moon, nearby asteroids, and Mars, hoping to learn about how the water we've discovered on other worlds could sustain a human colony. Every year since 2006, an international group of scientists and entrepreneurs holds a meeting in Washington State to plan for a space elevator that they hope to build in the next few decades. Such a project would allow people to leave Earth's gravity while using a minimum of energy, thus making travel off-world more economically feasible (and less environmentally damaging) than with rockets. Other groups are figuring out ways to reengineer our entire planet to slow the release of greenhouse gases and grow enough food for our booming population.

These projects, designed to improve cities on Earth while paving the way for life on other worlds, are just a few examples of how humans are getting ready for the inevitable mega disasters that await us. They are also powerful evidence that we want to help each other survive.

Human beings also have one survival skill that we've yet to find in creatures around us. We can pass on stories of how to cope with disaster and make it easier for the next group who confronts it. Our tales of survival pass over borders and travel through time from one generation to the next. Humans are creatures of culture as well as nature. Our stories can offer us hope that we'll make it through unimaginable troubles to come. And they can inspire scientific research about how we'll do it. Call them tales of pragmatic optimism.

This book is full of such tales—stories about people whose pragmatic optimism could one day save the world. Scientists, philosophers, writers, engineers, doctors, astronauts, and ordinary people are working tirelessly on world-changing projects, assuming that one day our lives can be saved on a massive scale. As their work comes to fruition, our world becomes a very different, more livable place.

If humans are going to make it in the long term, and preserve our planet along with us, we need to accept that change is the status quo. To survive this far, we've had to change dramatically over time, and we'll have to change even more—remolding our world, our cities, and even our bodies. This book is going to show you how we'll do it. After all, the only reason we're here today is because thousands of generations of our ancestors did it already, to make our existence possible.

PART I A HISTORY OF MASS EXTINCTIONS

Earth's **Deadliest Events**

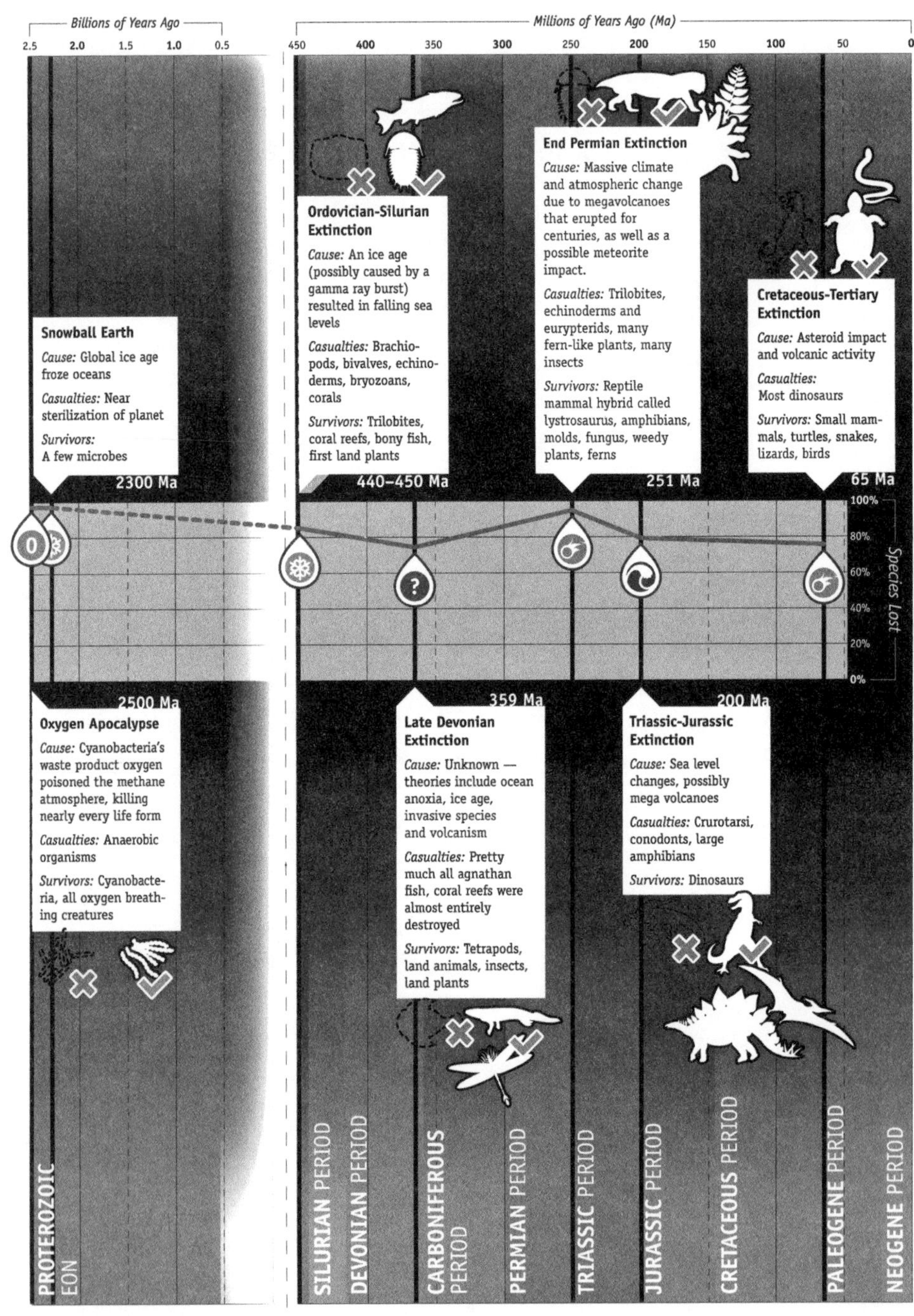

A timeline of mass extinctions, including Snowball Earth.

1. THE APOCALYPSE THAT BROUGHT US TO LIFE

IF YOU THINK that humans are destroying the planet in a way that's historically unprecedented, you're suffering from a species-level delusion of grandeur. We're not even the first creatures to pollute the Earth so much that other creatures go extinct. Weirdly, it turns out that's a good thing. If it hadn't been for a bunch of upstart microbes causing an environmental apocalypse over 2 billion years ago, human beings and our ancestors never would have evolved. Indeed, Earth's history is full of apocalyptic scenarios where mass death leads to new kinds of life. To appreciate how these strange catastrophes work, we'll have to travel back in time to our planet's beginnings.

The Proterozoic Eon (2.5 billion–540 million years ago): Oxygen Apocalypse

Earth is roughly 4.5 billion years old, and for most of its life the atmosphere would have been noxious for humans and all the creatures who live here now. Vast acidic oceans roiled in what today's environmental scientists would call an extreme greenhouse climate: the air was superheated and filled with methane and carbon. Our planet's surface, now covered in cool water and crusty soil, was bubbling with magma. The solar system had formed relatively recently, and chunks of rock hurtled between the young planets—often landing on them with fiery explosions. (One such impact on Earth was so enormous, and threw off so much debris, that it formed the Moon.) It was on this poisonous, inhospitable world that life began.

About 2.5 billion years ago, early in an eon that geologists call the Proterozoic, a few hardy microbes who could breathe in this environment drifted to the surface of the oceans. These microbes, called cyanobacteria (or blue-green algae), knit themselves into wrinkled mats of vegetation. They looked like black, frothy coats of slime on the water, trailing long, feathery tendrils beneath the waves. All that remains of this primordial ooze are enigmatic fossils that hide inside a distinctive type of ancient, spherical rock called a stromatolite. If you slice a stromatolite down the middle, you'll see thin, dark lines curving across its inner surface like the whorls in a fingerprint—these are all that remain of those algal mats. Only a few people in the world would recognize them as the traces of impossibly old life that they are, and Roger Summons is one of them. He's a geobiologist at the Massachusetts Institute of Technology who has spent decades studying the origins of life on Earth, as well as the extinction events that wipe it out.

An Australian with a dry sense of humor, Summons has an office you can only reach by walking through his lab, a big, airy room full of tanks of hydrogen and bulky mass spectrometers that look like old-school Xerox machines covered in tubes. When I visited him to talk about ancient Earth, he plucked some slices of stromatolite from the top of a filing cabinet to show me the traces of algae that spidered across their surfaces. "This one is eight hundred million years old, and this one is two-point-four billion," he said, pointing at each ragged half sphere of rock. "Oh, and this one is probably three billion years old, but it's a crap sample."

Even with a "crap sample," Summons can pin a date on the fossils of creatures who lived more than 2 billion years ago by examining the sediments that have preserved them. In his lab, researchers grind up ancient rocks, subjecting them to vacuum, freezing, lasers, and a strong magnetic field before running them through the mass spectrometers. At that point, often nothing remains of a stromatolite but ionized gas. And that's exactly what mass spectrometers need to decode the atoms in each sample. Atoms in minerals decay at a fixed rate, and reading the state of a rock's atoms can tell scientists how long it has been since it formed. Geologists don't put fossils themselves beneath the laser. They use machines like the ones in Summons's lab to figure out the ages of the rocks next to the fossils. Call it dating by association.

Knowing when the oldest stromatolites were created helps us date an event which changed Earth forever. The mats of algae that became stromatolites weren't just methane-loving scum. They were also filling the atmosphere with a gas that was deadly to them: oxygen. This is how the first environmental disaster on Earth began.

Just like plants today, ancient blue-green algae nourished themselves using photosynthesis, a molecular process that converts light and water into chemical energy. Cyanobacteria were the first organisms to evolve photosynthesis, and they did it by absorbing photons from sunlight and water molecules from the ocean. Water molecules are made up of three atoms—two hydrogen atoms and one oxygen atom (hence the chemical formula H_2O). To nourish themselves, the algae used photons to smash water molecules apart, taking the hydrogen to use as an energy source and releasing the oxygen molecules. This proved to be such a winning adaptation to Earth's primordial environment that cyanobacteria spread across the face of the planet, eventually exhaling enough oxygen to set off a cascade of chemical processes that leached methane and other greenhouse gases from the atmosphere. The dominant form of life on Earth ultimately released so much oxygen that it changed the climate dramatically, soon extinguishing most of the life-forms that thrived in a carbon-rich atmosphere. Today we worry that cow farts are destroying the environment with methane; back in the Proterozoic, it's certain that algae farts ruined it with oxygen.

Greenhouse Becomes Icehouse (and Vice Versa)

What happened after the rise of oxygen was an event shrouded in mystery until the late 1980s, when a Caltech geologist named Joe Kirschvink asked his student Dawn Sumner to research a rock whose existence seemed to be impossible—at least, given the prevailing theories about early Earth. Found near the equator, the rock's surface was scored with marks that suggested it had once been scraped by the weight of a slow-moving glacier. In a short paper that eventually revolutionized geologists' understanding of climate change, Kirschvink suggested that this rock offered a window on a late-Proterozoic phenomenon he called Snowball Earth.

Snowball Earth is what happens when our planet's climate enters a very extreme "icehouse" state, the opposite of a greenhouse. A carbon-rich atmosphere can heat our climate up into a sweltering greenhouse, but an oxygen-rich atmosphere cools it down and causes what's called an icehouse. Throughout its life, the planet has vacillated between greenhouses and icehouses as part of a geological process called the carbon cycle. Put in the simplest possible terms, a greenhouse happens when carbon is free in the air, and an icehouse occurs when carbon has been locked down or sequestered in the oceans and rocks. During an icehouse, ice collects at the poles, sometimes creeping down into lower latitudes during an ice age. But our recent ice ages were nothing compared with Snowball Earth.

Two billion years ago the sun was dimmer than it is today. As more and more cyanobacteria pumped out oxygen, the whole place began to cool down. Because the sun was a relatively weak heat source, this effect was magnified into a "runaway icehouse." Ice from the poles began to spread outward, solidifying the top layer of the oceans and burying the land beneath vast, frozen sheets. The more ice that formed, the more it reflected sunlight—lowering the planet's temperature further. Finally, ice stretched from the poles nearly all the way to the equator, pulverizing rocks beneath its weight. If you looked at Earth from space at that time, you'd have seen a slushy white ball, its circumference banded by a narrow equatorial ocean of algae-infested sludge. At that moment in geological history, our planet resembled Saturn's icy moon Europa. It was an alien world called Snowball Earth.

I visited Kirschvink at the California Institute of Technology to find out what happened next. In the basement of the geology building, his generously sized desk was piled with fossils, family photographs, papers, and his prized possession, a cheap plastic vuvuzela from South Africa. "This is real!" he enthused, gesturing at the instrument whose droning sound annoyed and delighted audiences during the 2010 World Cup. Kirschvink lit up when he talked about the provenance of objects, whether pop culture ephemera or 3-billion-year-old fossils. Maybe it was his off-kilter imagination that allowed him to look for environmental patterns in Earth's history that nobody had thought possible.

Kirschvink believes that there may have been as many as three snowball phases on Earth. "It was the longest, weirdest perturbation in the carbon cycle," Kirschvink said. "And my explanation for it is simple. It's the time between when the biosphere learned to make atmospheric oxygen and the time when everybody else learned to breathe it and use it." Without any creatures around to breathe oxygen, the cyanobacteria likely created an atmosphere far more oxygenated than any we've ever known.

For 1.5 billion years after cyanobacteria evolved, Earth's biosphere was in chaos. At least two more snowballs crept across the face of the planet, followed by intensely hot greenhouse conditions caused when volcanoes pumped carbon back into the air. Meanwhile, microbes were slowly learning to use oxygen to their advantage. A new kind of cell called a eukaryote began to populate the seas. Unlike cyanobacteria, which are basically just genetic material contained inside a membrane, a eukaryotic cell contains a nucleus packed with DNA as well as tiny organs called, appropriately enough, organelles. One of those organelles, called a mitochondrion, could turn free oxygen and other nutrients into energy. At last, Earth was inhabited by oxygen-breathers. The planet we know today was taking shape.

While the eukaryotes got busy swapping genetic material and sucking oxygen from the air, the old methane-breathers were dying out. A few migrated to the sea floor, finding niches near superheated volcanic vents where they could live in the remaining fragments of a once-global methane ecosystem. But the rest went extinct. It was the most extreme form of atmospheric pollution in Earth's history, soon killing off almost every form of life that couldn't breathe oxygen.

By "soon," I mean within a billion years, or possibly 2 billion—a period of time that's almost impossible to wrap our minds around. Still, that is the timescale required to understand Earth's environmental transformations. Many of the catastrophic changes we'll discuss over the next few chapters took millions of years to unfold. To geologists, we are all living in fast motion, our lives so short that it's usually impossible for us to personally experience environmental change. Often, these scientists will contrast "human-scale" time with what they clearly view as real time, or time that unfolds on a planetary scale.

One of our most incredible accomplishments as a species, however, is an ability to think beyond our own life spans. We may not live in geologic time, but we can know it. And the more we learn about our planet's past, the more it seems that Earth has been many different planets with dramatically different climates and ecosystems. This idea offers a much broader perspective than what you find in the work of environmentalists like Bill McKibben, who argues in his book *Eaarth* that humans have burned so much fossil fuel that we're turning our planet into something fundamentally different (requiring the new name Eaarth). In that book and elsewhere, he laments the loss of "nature," by which he means the ecosystems that existed on Earth before human meddling. But before humans took center stage on Earth, there were many permutations of nature. Climate disasters were the norm. Indeed, the only way Earth could ever transform enough to merit a new name like Eaarth would be if the planet's environment suddenly stopped changing.

Undeniably, our planet is undergoing potentially deadly environmental changes today. But it's incorrect to say that this is the first or even the worst time it's happened. For the creatures who perished during the Proterozoic, and other periods we'll learn about in the coming chapters, McKibben's ideal of nature would be deadly. Over the course of its history, Earth has always vacillated between a carbon-rich greenhouse and its opposite, the oxygen-rich icehouse where humanity is more comfortable. We're simply the first species on Earth to figure out how this climate cycle works, and to realize that our survival depends on preventing the next environmental shift.

Defining Mass Extinction

As bad as the oxygen apocalypse was, neither Kirschvink nor the geobiologist Roger Summons would call it a mass extinction. So how can an entire world full of life go extinct without it being a mass extinction? This brings us to the question of what mass extinction really is. In a remarkable paper published in *Nature* in the spring of 2011, a group of biologists from across North and South America exhaustively summed up all the data available from the fossil record and present-day extinctions and came up with a clear definition. They agreed that mass extinctions on Earth can be defined as events in which 75 percent or more species go extinct in less than 2 million years. The oxygen apocalypse didn't happen fast enough to qualify.

The statistician and paleontologist Charles Marshall, a coauthor on that *Nature* paper, warns that the definition of "mass extinction" is highly contextual and slippery. Sitting with his back to an enormous window overlooking the UC Berkeley campus, Marshall told me that the key to understanding mass extinction always begins with a calculation of what researchers call the "background extinction rate." Species naturally pass into extinction all the time, at a rate of about 1.8 extinctions per million species every year. On top of that, there are also natural cycles of elevated extinction rates that fall roughly every 62 million years in the fossil record. So just because a bunch of creatures are going extinct, even in numbers above the background extinction rate, doesn't mean you're looking at a mass extinction. The only time you're really seeing a mass extinction, Marshall said, is when "you see a big spike sticking out of the background distribution." While on Earth those big spikes tend to be times when 75 percent or more species go extinct, it's all relative. "You could imagine a planet where the biggest spikes sat at thirty percent," Marshall speculated. "On that planet, thirty percent of species dying out would constitute a mass extinction."

There are some ways that the fossil record can trick us into seeing a mass extinction where there isn't one. Take, for example, the bombs at Hiroshima and Nagasaki. The rates of death were high, but they were low in terms of the world's population. If we looked at these atomic bomb strikes in the fossil record, it might appear that there had been a mass extinction, but that's because we'd be mistaking the rates in one local area for a global phenomenon. When geologists study mass extinction in the fossil record, they constantly have to ask themselves whether the extinctions they're seeing are a statistical anomaly like Hiroshima, or something more widespread. Mass extinction is not an absolute idea, and to measure it we have to prove that the extinctions aren't just localized. Plus, we have to compare the rate of death to the normal background extinction rate.

Still, the oxygen apocalypse does resemble a mass extinction in one way. It ushered in a completely different world, populated by an entirely new set of life-forms. It gave rise to the atmosphere that allowed life as we know it to develop. The change was so dramatic, said Marshall, that "you're measuring less by magnitude and more by the idea of a world changed forever." In every mass extinction, the world is changed forever—but over a short, terrifying two million years, rather than a slow billion. In the next few chapters, we're going to see exactly what that looks like.

2. TWO WAYS TO GO EXTINCT

NEARLY TWO CENTURIES ago, scientists trying to learn about Earth's history visited England's famous "white cliffs of Dover," where the wind and water had eaten away at the land's edge, revealing rocks unseen for millennia. There, these early geologists discovered that the cliffs weren't just big, crumbling slabs of chalk. They were actually formed from distinct layers of rock, each containing very different sets of fossils, providing a chronological record of how the land was built up from mineral deposits and ecosystems through the ages.

Each geological period is named for one of these rocky layers. They're chunks of frozen time, identified by their unique combinations of lifeforms and mineral deposits. Generally, fossils change dramatically from layer to layer because there has been an extinction event. Though only five of these demarcations qualify as mass extinctions, there have been dozens of smaller extinction events where, say, 20 or 30 percent of all species die out. You might say that geological time is measured in extinctions. But if you were to visit Dover, and allowed your eyes to wander up the cliff face, you'd also see layer after layer of evidence that life always emerges from mass death.

The more we learn about these layers, the more it seems that there are two basic causes that can set a mass extinction in motion. The first is an unexpected calamity from the inanimate physical world, often taking the form of fatal climate patterns, megavolcanoes, or even debris from explosions in space. As for the second, as we learned in the last chapter from the cyanobacteria that poisoned the planet, biological life can transform the physical world so much that extinctions are inevitable. Of course, many extinctions are a combination of the two causes–one often leads to the other.

To see examples of both, we'll journey back hundreds of millions of years to the first two mass extinctions that gripped the Earth. The Ordovician (beginning roughly 490 million years ago) and Devonian (beginning roughly 415 million years ago) were both periods when life was exploding with unprecedented diversity. And both ended in holocausts. The Ordovician was scourged by natural disasters from Earth and space; the Devonian was choked to death by invasive species that turned the planet into an environmental monoculture.

The Ordovician Period (490 Million–445 Million Years Ago): How the Appalachians Destroyed the World

Before the fecund Ordovician period, the seas had stopped looking like goo-covered murk and flowered into underwater forests full of aquatic plants, shellfish, coral, and lobsterlike arthropods called trilobites. New species were evolving at a rapid clip. It was a greenhouse world, with carbon dioxide levels in the air at fifteen times higher than they are today. But a warm, high-carbon climate was exactly what those Ordovician plants and animals needed.

Peter M. Sheehan, a geologist with the Milwaukee Public Museum, describes the Ordovician as having "the largest tropical shelf area in Earth's history." Put another way, it was a world of sultry beaches. Earth blossomed into this tropical paradise partly due to the climate, and partly due to continental drift, the process by which massive plates of the Earth's crust slowly move around on top of the planet's superheated molten layers. Lava gushing from underwater volcanoes applied so much pressure to the Earth's crust that it pushed all the continents together, into the

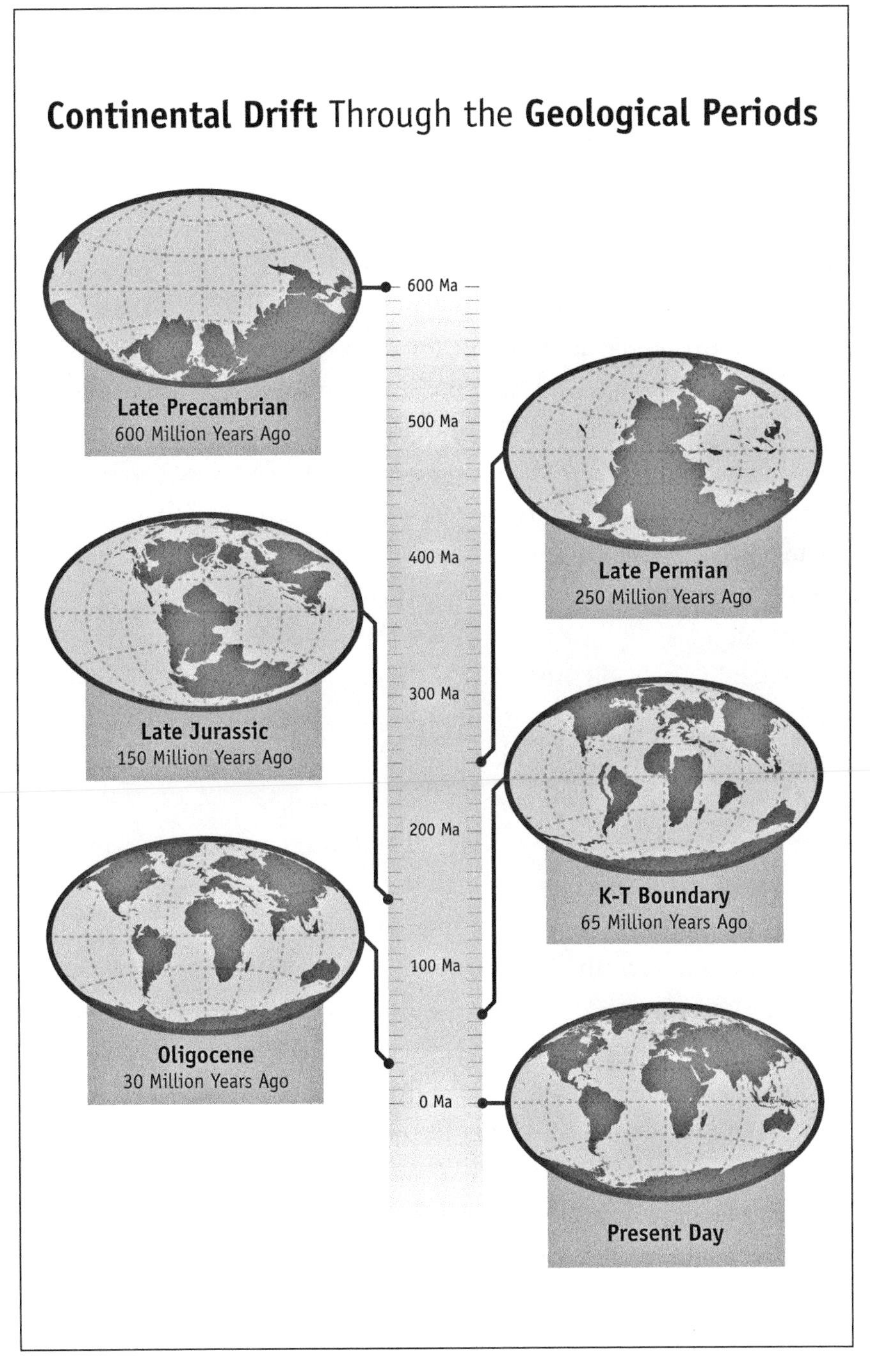

Maps of the continents during different geological eras.

low latitudes of the warm southern hemisphere. Slowly drifting over the South Pole was a supercontinent called Gondwana, made up of land that became, among other places, Africa, South America, and Australia. Its balmy, world-wrapping coastline teemed with life.

Ordovician life was confined almost entirely to the oceans, though a few plants spread to the land. Trilobites scuttled into many different territories, evolving into a range of species: some became swimmers, while others wandered the floors of the shallow seas, developing sharp, defensive spines or shovel-shaped heads for rooting food out of the sediment. Shelled creatures and sea stars attached themselves to enormous coral reefs, and strange colony animals called graptolites built complicated, beehive-like structures out of proteins secreted from their bodies. Their hives, which looked like thorny, interconnected tubes, floated beneath the ocean surface while the graptolites poked their feathery heads out and snarfed up plankton.

The ancestors of sharks prowled the waters and fed on everything that moved (and some things that didn't). Joining the sharks were jawless fish called agnathans, whose soft mouth slits and heads were covered in bony plates that probably looked like turtle shells. These armored fish were the first vertebrates, or animals with internal skeletons like we have. Plus, there were thousands of new kinds of plankton evolving all the time, creating an abundant food source for all the multicelled newcomers looking for easy-to-reach food floating through the waters.

But over a few hundred thousand years, over 80 percent of the species in the Ordovician coastal waters would go extinct.

We can place part of the blame for the slaughter on the Appalachian Mountains, a gently curving spine of peaks that stretched from Canada's Newfoundland down to Alabama in the southern United States. These mountains were formed during the Ordovician when a smashup between two continental plates pushed ancient volcanic rock into jagged peaks above the continent. Almost immediately, rain and wind began eroding the soft, dark rock. The newly formed mountains ran with thick slurries of water and mud, which turned into rivers that picked up even more soil on their way to the seas. This natural process, called weathering, is actually one of the most powerful ways to change our planet's atmosphere. As exposed earth crumbles beneath the weather's onslaught, tiny rocks pull carbon dioxide from the air and take it with them into sediments deep beneath the sea. Sliding into the sea along with all that carbon was the Ordovician's warm, life-nourishing climate.

Seth Young, a geology research associate at Indiana University, observed, "We are seeing a mechanism that changed a greenhouse state to an icehouse state, and it's linked to the weathering of these unique volcanic rocks." The Ordovician Appalachians weathered so rapidly, in fact, that they were worn down to a flat plain within a few hundred million years. The Appalachians we know today are the result of a second tectonic-plate smashup, which raised a new set of mountains about 65 million years ago. Washing carbon out of the atmosphere sounds like a good dream in our fossil-fueled times, but it was the worst thing that could happen in the Ordovician. Without greenhouse gases to keep the planet warm, disaster struck in the form of the fastest glaciation in the planet's history. About 450 million years ago, ice caps began to spread outward from the poles. Gondwana and its hot, humid shorelines were at ground zero of the ice apocalypse.

As the glaciers grew, they locked up liquid water and lowered sea levels dramatically, drying out the lush coastal areas beloved by corals, graptolites, and shelled creatures. Most affected were stationary animals like the shellfish in a coral reef, which remain anchored in place for most of their lives. Because they couldn't move, they died with their habitats. In all, Peter Sheehan estimates that about 85 percent of marine species died over a million years as massive ice sheets sucked the liquid out of their environments. Not all the Ordovician species died at once. There were two "extinction pulses," as geologists put it. The first came when ice abruptly destroyed sea life. The second came when the ice melted just as suddenly as it had come, causing sea currents to slow and stagnate. Fewer currents meant that less oxygen was churned into the water and vast "dead zones" of anoxic (low-oxygen) water suffocated life throughout the oceans. First came ice, then came stagnation. Together, they created a mass extinction.

Despite what we know about the Appalachian Mountains, the wholesale slaughter at the end of the Ordovician remains a mystery. We understand why rapid freezing and thawing would kill so many life-forms. But typical ice ages are millions of years in the making, and this one lasted for less than a million, making it ridiculously rapid in geological time. Could weathering alone have caused the rapid glaciation in the first place? Probably not. It's possible that the ice formation was hastened by invisible rays from space.

Cosmic Rays of Death

Adrian Melott, a professor of physics and astronomy at the University of Kansas, has long been fascinated by a weird fact about mass extinctions. They seem to fall roughly every 63 million years. Trying to explain why this might be, he stumbled upon one possible explanation for the swiftness of the Ordovician ice age. It has to do with the motion of our star through the swirling galactic disk of the Milky Way.

Every star in the galaxy has an orbit around the edges of the galactic disk. As our sun makes its vast circuit around the Milky Way, it bobs up and down, floating above or below the galaxy's flat plane about every 60 million years. When it does this, our solar system brushes the edge of the protective magnetic field that envelops the galaxy, deflecting dangerous cosmic rays zooming through deep space (on a smaller scale, the Earth's magnetic field protects us from these same particles). Cosmic radiation could help explain why extinction events are more likely to happen every 63 million years or so.

Cosmic rays are highly energetic subatomic particles that have been bouncing around in deep space since the early days of the universe. They can shoot right through a living creature's body, damaging its DNA and causing cancer. And these particles aren't much kinder to the molecules that make up Earth's atmosphere. Cosmic rays can damage the ozone layer, which leaves the planet more vulnerable to deadly radiation. Melott hypothesizes that cosmic-ray bombardment could also whip up a thick cloud layer in the atmosphere, lowering temperatures and helping the ice caps to form more quickly.

As the planet cooled, extinctions would have been worsened by radioactivity hitting the planet's surface. "At this point we're thinking that... the radiation dose for organisms on the surface of the earth, or in the top kilometer of ocean water, could be very large. This causes cancer and mutations." Melott paused, as if imagining a planet with gray skies racked by cancers and catastrophic erosion. Then he chuckled. "Or, you know, it could lead to giant ants that rampage across the Earth."

His joke about the 1950s atomic monster movie *Them!*, featuring giant ants that take up residence in the sewers of Los Angeles, underscores the degree to which he thinks of his work as speculative. Cosmic rays, he conceded, were only one part of the problem that animals faced at the end of the Ordovician. "The analogy I like to give is that it's like you have the flu and then you get shot. Cosmic-ray stress is like the flu." But other factors—the bullet in Melott's analogy—need to be in play. And these would likely be the volcanic activity that led to the uplift of the Appalachians, the weathering that flattened them, and the resurgence of volcanoes that shut down the Ordovician ice age as quickly as it began.

The Ordovician ended with an extremely rapid version of what happened during the snowball phases of Earth's history. A swiftly changing climate, vacillating between icehouse and greenhouse, made it impossible for most species to survive. Because those deadly climate shifts happened so fast, geologists have dubbed this horrific period the Ordovician mass extinction, marking the first time our planet witnessed the deaths of so many species at once.

The Devonian Period (415 Million–358 Million Years Ago): Invasive Species

By the time the planet's temperatures had stabilized, the Ordovician biosphere was gone forever. A few survivors remained, like the hardy trilobites. But for the most part, new animals and plants evolved to rule the seas, and a few creatures even crept up on land. Life diversified and flourished for 100 million years, a fairly long time even for a geologist. Consider that modern humans evolved only about 200,000 years ago, and you have an idea how many species evolved and died out during the 100 million years before the planet suffered its next mass extinction. Our next rendezvous with mega death came during the Devonian period. This time there were no dramatic claws of ice, cosmic rays, or greenhouse extremes–but that's because this was the first mass extinction caused by life itself. By the end of the Devonian, 50 percent of marine genera (groups of species) and an estimated 75 percent of species were dead. Oddly, these species died out at an ordinary rate, probably no higher than the typical background extinction levels. So why is this even considered a mass extinction at all? Because almost no new species evolved to take the extinct ones' places for as many as 25 million years. It was mass extinction by attrition.

Scientists call this phenomenon a "depression in speciation," meaning a low point in the evolution of new species. If you had been floating around for thousands of years in one of the Earth's oceans during the late Devonian, about 374 million years ago, you wouldn't see corpses piling up. Nor would you see vast stretches of lifeless water as you would have during the late Ordovician. Instead, you'd see the same species slowly spreading everywhere, darting around in enormous coral reefs that were ten times more expansive than the ones we have today. There weren't fewer lifeforms during this mass extinction. There were just fewer kinds of them.

How did invasive species destroy the planet? It all had to do with the period's peculiar ocean ecosystems. The massive sea creatures of the Devonian earned the period its nickname, the Age of Fishes. The eminent geologist Donald Canfield conducted a study of the atmosphere during this period, after which he and his colleagues concluded that the Devonian oceans contained a high amount of oxygen, which allowed the period's enormous animals to evolve. A group of hardy armored fish called placoderms vied with sharks to become the ocean's most forbidding predators. Placoderms grew up to 36 feet long and had faces entirely covered in armor; they were also among the first creatures to develop jaws. (Sharks won the scary toothed predator contest in the end, though–placoderms went extinct.) Reefs made from algae and early sponges–all species lost to this extinction–were dramatically unlike the coral-dominated reefs we know today. The ocean floors crawled with ammonites, which looked something like octopuses with spiraled shells.

Watery habitats were everywhere—even on the continents. Enormous tropical inland seas dominated the landmass that later became North America. Most of the Midwest and the central United States were fully submerged, which is why paleontologists today find some of the best fossilized fish in the middle of the Midwest's rolling prairies, which are about as far from the coast as you can get. Yet by the end of the Devonian, almost none of the gigantic armored fish and swarms of tentacled ammonites were left. What happened?

One paleontologist, Ohio University's Alycia Stigall, has a theory that could explain why life during this period went from diverse to homogenous. She believes that invasive species took over the world's oceans and inland seas, the same way cockroaches, kudzu, rats, and humans have spread across the globe today.

Stigall lives in Ohio, at the bottom of what was once a shallow Devonian sea. In fact, the vista from her windows is the former seafloor of an inland ocean hit particularly hard by the mass extinction. "We don't have a good modern analogue for these types of oceans," she said. They are a completely vanished ecosystem, though she imagines they might have been something like Hudson Bay. At the end of the Devonian, it's likely that sea levels were high, pooling several inland oceans together. Earthquakes thrust new mountains from the land, which also brought previously separated ecosystems into contact with each other.

Many highly adaptable or generalist species began invading new watery territories. That meant they were competing with the local specialist species, like trilobites, for food. A specialist species requires very specific temperatures or food sources. They couldn't cruise all over the Devonian oceans eating anything that came into view the way sharks could. So when invasive species came into their territories and stole their food, the specialists had nowhere to go. Their populations dwindled and they went extinct. By the end of the Devonian mass extinction, the planet was covered with giant, homogenous inland oceans where you'd see the same generalist species no matter where you looked. And out of those circumstances, Stigall believes, you had the makings of a mass extinction.

Though Stigall's account is only one of many theories about mass extinction in the Devonian, her claims are backed up by evidence that many of the creatures who survived the period were generalists like sharks—creatures who could live anywhere and eat almost anything. Another survivor was the humble crinoid, a starfish-like creature with several feeding arms surrounding its mouth that make it look a little like the "face-hugger" stage of the creature in *Alien*. The crinoids went through a floating larval stage that allowed them to drift into many new environments before attaching to the ground and feeding on the abundant plankton in the water.

Still, homogenous ecosystems like the ones in the Devonian would have left all life-forms primed for disaster. Generalist species may be hardy, but they also share the same vulnerabilities. Say, for example, a drought hits an ecosystem in the Midwest. If there is only one type of wheat species, and it can't deal with higher temperatures or less moisture, a short-term climatic change could kill off every blade of wheat in a given region. Without a diverse range of grain species, which might have different moisture tolerances, drought slays all the wheat. This in turn kills the animals who feed on that wheat, whose deaths leave predators hungry, too. Soon, you have multiple extinctions because the whole food chain has been ravaged. "The more we cull diversity, the more we are vulnerable to extinction," Stigall concluded. It's very possible that this is how the Devonian came to a close, with only a few invasive species scraping by until speciation, or new species evolution, made the ecosystem diverse again.

For Stigall, there's a lesson in the fossils that surround her home, remnants of an inland sea unlike anything that exists now. "We've got the same problem with invasive species today," she said. "It's not caused by sea levels, but by humans, because we like to move things around." She described how invasive species like pigeons and rats, as well as some trees and grasses, have gone from a few local regions to expand across the Earth. If this trend continues, she predicted that "long term, what you expect is a huge decimation of total biodiversity." We could be on our way back to the late Devonian, in the early stages of a mass extinction that begins with a depression in speciation and ends with deadly homogenization.

Sometimes, the urge to live by expanding into as many territories as possible can backfire. As the invasive species of the Devonian reveal, life does not always beget more life. Some ways of living can actually kill just as handily as climate change and radiation.

As devastating as the Ordovician and Devonian mass extinctions were, they were nothing compared to what ravaged the planet about 75 million years later. The "Great Dying," as it is known among geologists, was the worst period of mass death the Earth has ever known. It's very likely that this event had no single cause; it was set off by a combination of disasters from the physical world and the biological.

3. THE GREAT DYING

THE BERKELEY GEOCHRONOLOGY CENTER, a lab devoted to studying the ancient ages of Earth, is located on a pleasant tree-lined ridge overlooking UC Berkeley. Somewhat puzzlingly, it shares a building with the Church Divinity School of the Pacific. While seminary students strolled by in the hall outside, I met Paul Renne, the center's head geologist, a big, jovial man in a T-shirt. As he walked me through a warren of labs—full of the lasers and mass spectrometers that I'd come to expect in such places—I realized there was a peculiar kind of symmetry to the Geochronology Center's tenancy arrangement with the Divinity School. After all, I had come to ask Renne about the closest thing to a total apocalypse the planet has witnessed.

The Permian Period (299 Million–251 Million Years Ago): Life in the Time of Megavolcanoes

Two hundred fifty million years ago, at the end of the Permian period, Earth spent thousands of years dying. At the end of those millennia of carnage, almost 95 percent of the species on the planet were dead. It was the worst mass extinction in our planet's history, earning it the moniker "the Great Dying."

The first phase of the mass extinction was caused by a disaster that has left an indelible and easily deciphered mark upon the Earth. If you visit the vast area known as the Siberian Traps today, you'll find a beautiful, hilly terrain covered in short grasses. But 250 million years ago, the region was drowning under liquid rock spewing from the ragged mouth of a megavolcano. As its name implies, a megavolcano is far more powerful than your typical lava-filled mountain. Renne and other geologists estimate that as much as 2.7 million square kilometers of basaltic lava swept across the land in a fiery deluge. Almost a million square kilometers of the hardened basalt rock still remains here, smoothed by erosion into plateaus and valleys. It's unclear whether this almost unimaginable ocean of lava was unleashed by one or two enormous eruptions, or a single, ongoing eruption that lasted for centuries.

But the Great Dying wasn't caused by flaming tides of death. Volcanic eruptions on a large scale release a lot of gases, including greenhouse gases like carbon dioxide and methane. Jonathan Payne, a geologist at Stanford, estimates that the eruptions unleashed 13,000 to 43,000 gigatons (a gigaton is 1 billion tons) of carbon into the atmosphere. As if that wasn't enough, they also released highly reflective sulfur particles that remained suspended in the atmosphere, scattering light away and cooling the climate very rapidly. The culprit responsible for the Great Dying was climate change.

Ironically, the roiling fires from this Siberian megavolcano may have caused a brief ice age. As glaciation locked coastal waters into ice sheets, the sea level dropped, and another source of greenhouse gas was unleashed. It's possible that the water dipped low enough to expose methane clathrates, huge deposits of frozen methane that cling to the edges of continental shelves deep beneath the ocean. The clathrates melted and released ancient methane, a powerful greenhouse gas. As quickly as it began, the Permian ice age would have ended with a more intense greenhouse than before. These radical transformations in the atmosphere and climate made it impossible for most creatures to survive. Food sources dwindled. Species upon species died out.

It was an ugly ending for the Permian, which had been a time of rapid

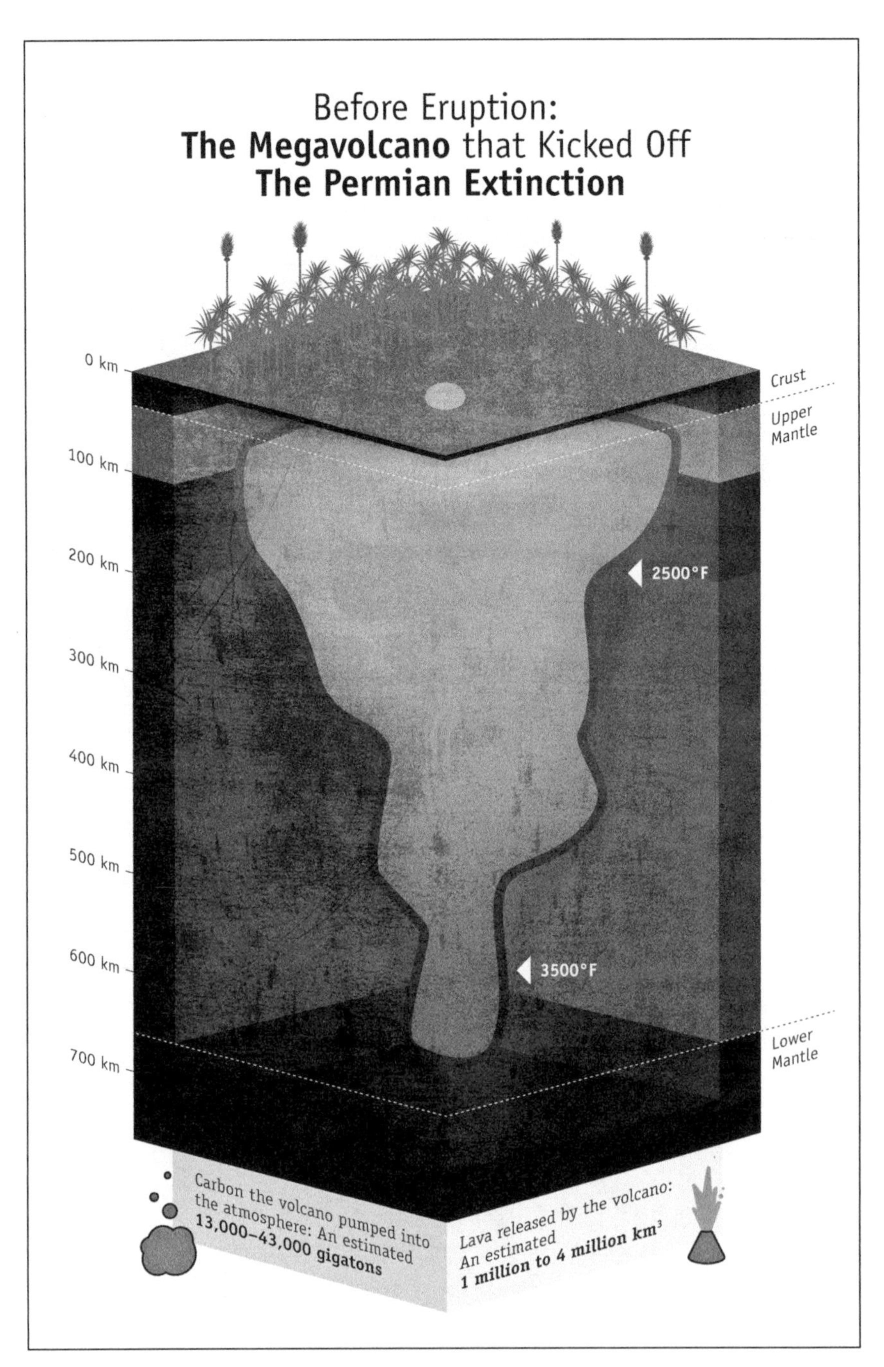

A cutaway view of the megavolcano in Siberia that led to the Permian mass extinction.

animal evolution on land. When the megavolcano began erupting, the earliest ancestors of today's mammals were walking the Earth. Gingkos and conifers covered the coasts in forests, while seeded ferns evolved, uncurling their leafy fronds beneath tall pines. Mammal-reptile hybrids called synapsids roamed the land, some looking like giant lizards, and some like small rhinos. One of them, the enormous, dragon-like predator dimetrodon, had a tall sail attached to its back like a bony fin, and was such a badass hunter that paleontologists believe it may have fed upon sharks. These creatures all thudded around on the same continent because plate tectonics had finally pushed the planet's landmasses together into one enormous continent called Pangaea, which stretched from pole to pole. A globe-wrapping ocean called Panthalassa teemed with sea creatures, from tiny single-celled organisms to corals and large fish.

These new forms of life, the forerunners of so many animals and plants we take for granted today, almost didn't make it. What was especially unusual about the Permian mass extinction was that it took out nearly every form of life. Unlike in other mass extinctions, which sometimes hit sea creatures but not land creatures, or animals but not plants, this extinction was absolute. As many species were lost at sea as on land. When the megavolcano pumped carbon into the atmosphere, a lot of that got dissolved into the oceans. The water grew warmer, which destroyed the habitats of shellfish, who are sensitive to temperature changes. It also grew more acidic. The shells of shellfish are made of calcium carbonate, which dissolves in acid. Many sea creatures didn't survive simply because their offspring couldn't form shells in a highly acidic ocean environment.

Meanwhile, on land, so many trees and plants died that the continent's surface was "denuded," as Payne put it. The result was shockingly rapid weathering. As acid-tinged rain poured from the sky, followed by hot winds, more soil poured into the oceans, further raising the levels of carbon and acid. Vast areas of the coastal seas became anoxic dead zones—regions completely purged of oxygen. With oxygen supplies low in the water, large fish could not survive, especially ones that lived close to the deeply damaged ocean's surface.

Even insects, which generally survive everything, suffered extinctions.

An estimated 9 out of 10 marine species and 7 out of 10 land species went extinct. Across the planet, carbon levels suddenly skyrocket in rocks from this period examined by Renne and his colleagues, which suggests that the dead bodies of plants and animals were quite literally piling up on land and at sea. As the plants rotted, they released even more carbon into the environment. The devastation was so complete that we see a "coal gap" in the layers of rock left behind from this era. Plant life, which decays into coal, was so sparse in the 10 million years following the end of the Permian that none of the fossil fuel could form.

The planet had already endured ice ages, greenhouses, cosmic rays, and speciation depression. But only in the Permian mass extinction were almost 95 percent of all species cut down. And it happened in just 100 thousand years—the blink of an eye in geological time.

Slime World Survivors

Still, there were survivors. The Stanford geologist Payne showed me a rock that's a slice of geological time from this period, where a layer of ocean-floor sediment filled with tiny shells is topped by a black layer of what looks like pure sludge. It's easy to see that a diverse community of creatures was abruptly replaced by nothing but, well, slime. Payne and his colleagues have nicknamed this era Slime World, because the oceans were dominated by dark, oozing bacterial colonies, feasting on the dead bodies of their multicellular cousins.

On land, one of the great survivors was *Lystrosaurus*, an animal that managed to thrive. A heavy, clubfooted creature with a beaked snout and two tusk-like teeth, *Lystrosaurus* was a four-legged synapsid, or mammal-reptile hybrid. About the size of pigs, lystrosaurs were burrowing animals whose muscular hindquarters ended in short, wiggly tails. And they somehow managed to endure when even the precursors of the hardy cockroach were dying. They were herbivores, and their beaks probably allowed them to chomp on rough vegetation and dig for roots to eat.

For several million years after the end of the Permian, lystrosaurs were alone on a dead world. But they didn't cower or retreat. Instead, they

Lystrosaurus was one of the few land animals to survive the Permian mass extinction, and its progeny spread across the Southern Hemisphere during the early Triassic.

spread out as far as they could across the landmass that would one day fracture into the continents we know today. Their fossils have been found in Africa, Asia, and even Antarctica, which was a tropical region at the time. With no predators and no competition for their favorite foods, lystrosaurs could waddle anywhere they liked. They are, as far as we know, the only creatures ever to dominate our world so thoroughly: For millions of years, most four-legged land creatures were one type of lystrosaur or another.

Why did these creatures—our distant ancestors—survive when so many of their fellow creatures didn't? Theories abound. The Permian expert Mike Benton said it's possible that they were "just lucky." More likely, he added, they were well adapted for a world with depleted oxygen. They lived in underground tunnels, so they had a natural way to escape the heat and fire of the initial volcanic eruptions. Plus, the air they were used to breathing in their burrows was likely to be low in oxygen and full of dust–sort of like the air after carbon has been saturating it for a few centuries. Their barrel chests held lungs of a tremendous capacity, which meant more oxygen uptake. *Lystrosaurus* had the right respiratory system at the right time.

Over time, *Lystrosaurus*'s progeny repopulated the southern part of Pangaea, diverging into many subspecies. Their favored half of the supercontinent eventually broke off from the northern half and became its own continent, Gondwana (named after the southern Ordovician continent), packed with dinosaurs and proto-mammals. It took 30 million years for our planet to grow a robust ecosystem again, packed with predators and herbivores and a wide range of flora and fauna.

The Early Triassic Period (250 Million–220 Million Years Ago): Unraveling Food Webs

Those 30 million years of ecosystem struggles are their own story. Though every mass extinction unfolds differently, they all end when a new community of creatures has established itself–generally, a community that statistician Charles Marshall described as "completely different life-forms." After the Permian, during the early millennia of the Triassic period, new communities of completely different life-forms rose and fell with alarming regularity. A new ecosystem would come together only to collapse in a few million years. Then another ecosystem would arise. This mass extinction just wouldn't end.

Why did it take the planet so long to recover from the Great Dying? For answers, I visited Peter Roopnarine, a zoologist at the California Academy of Sciences who has a rather singular occupation among scientists. He's developed a computer program that simulates food webs, the complex interplay between predators and prey within an ecosystem. Using this program, Roopnarine studies why the worst part of mass extinctions

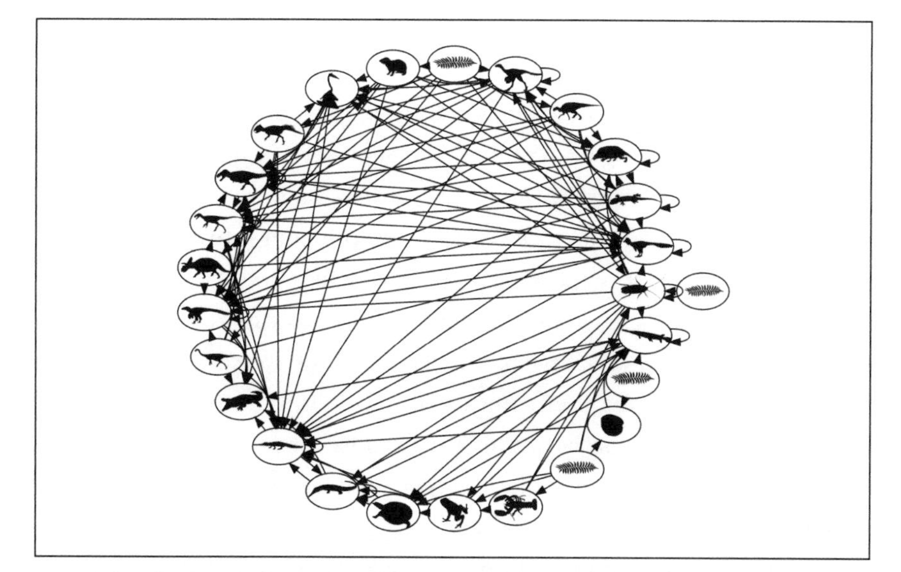

In this food web illustration created by Peter Roopnarine, the arrows between life-forms indicate who eats whom. This is a Cretaceous-era food web.

isn't necessarily the fire, or the eruptions. It's what comes afterwards, in the centuries of what scientists call "indirect extinctions" caused by food webs that are too unstable to support life.

The old computer game Wator offers a perfect example of a simple food web simulation. In it, red pixels stand in for sharks (predators) and green pixels for fish (prey) as they battle it out for supremacy of the sea. You can set a few simple parameters, such as how many sharks and fish there are to start, how often they breed, and how long it takes before they starve. Then you press "start" and watch generations unfold in seconds. When there are too many sharks, or the fish breed too slowly, the population of sharks eventually dwindles to zero and the waters of "Wator" become a sheet of uniform green. And that means you fail. What Wator reveals is that predators are as much at the mercy of prey as the reverse. Food webs can be knocked out of balance by life-forms at any point in the food chain.

Roopnarine's simulations are infinitely more complex than Wator, incorporating the smallest planktons to the largest predators, and everything in between. In them, he describes relationships between predators and prey that lived millions of years ago. And from these models, he's generated a theory about why the Triassic burned through so many food webs.

He began by coming up with a way to generate a realistic food web for species that no longer exist. He included every known form of life from the fossil record, and then he extrapolated predator–prey relationships based on what we know about how animals behave today. "You can't ever know exactly what a fossil animal was eating—you can't even know that with animals today," Roopnarine explained. "But we can use the body size, tooth shapes, and other things to decide who their prey might have been." Predators' body sizes are helpful because, obviously, a small predator will prefer small prey, while a larger predator might be a generalist who can eat creatures of many sizes.

There are always complications, Roopnarine admitted. Many species share the same potential prey or predators, and it's hard to know which species might have been generalists with many food sources, or specialists with just a few. But the beauty of using computers to simulate food webs is that you can go through as many iterations as you like, creating different worlds each time paleontologists discover more about a fossil predator's range or appetites. Plus, there are a few rules of thumb, including the fact that there are usually far more specialists than generalists. Once the ancient food web has been set up in his program, Roopnarine said, he can simulate food-web disturbances like the one in the early Triassic.

Based on what he's figured out so far, Roopnarine's theory is that a basic imbalance in early Triassic food webs led to millions of years of maimed ecosystems rising and collapsing in rapid succession. Initially, the problem was that so few creatures had survived the Permian mass extinction. Of the survivors, he said, "you have small carnivores and some seriously big, bad amphibians who are the precursors of crocodiles." Among herbivores, he said, lystrosaurs were the only game in town. The problem was that nobody around seemed to be eating *Lystrosaurus*, perhaps because they were the wrong size or in the wrong environments for most predators. In fact, the food web began to unravel because there were so many carnivores and very few prey.

Those "big, bad amphibians," known as crurotarsans, were in fierce competition with each other. With their huge, toothy mouths and muscular tails, they would have been deadly predators—and Roopnarine says that the competition between carnivores during the early Triassic was more intense than in any other food web he's looked at. The carnivores competed with each other so intensely for the tiny amount of available prey that they wound up driving each other to extinction over and over. New creatures would evolve, then get crushed out of existence. Only the herbivore *Lystrosaurus*, and eventually other herbivores, really recouped their losses. It took tens of millions of years before there were a small enough number of carnivores for food webs to stabilize.

Community Selection

This raises the question of what makes for a stable food web over the long term. And there's an easy answer. "Diversity," Roopnarine said firmly. A food web needs to be "robust," full of many kinds of carnivores, herbivores, and plants, in order to withstand an environment that can often hammer creatures with everything from volcanoes to drought and sea-level shifts. As long as there are many nodes in a food web, a healthy balance of predator and prey, you have a community of life-forms that can remain steady even when the environment wobbles.

"So does that suggest some communities are better than others when it comes to survival?" I asked.

Roopnarine offered a conspiratorial nod. "This can be controversial, but yes, you could say this is natural selection at the community level." Food webs don't compete the same way two species might because they don't exist next to each other, trying to eat the same things and live in the same caves. Instead, they compete with each other temporally, replacing each other in the same geographical places over time. To "win" the natural-selection game, a food web must outlast other food webs, remaining stable for as long as possible in the same place. Looked at from this perspective, you might consider all of Earth's geologic history a competition between food webs struggling to last through as many environmental disasters as possible, simply by retaining their robustness in the face of calamity.

Survival is never just a matter of one species being exceptionally adept. We only survive in the context of our food webs. And when a food web starts to unravel, the extinction of one creature will mean the "secondary extinctions" of others.

Roopnarine and his colleagues have run enough simulations of food-web collapse that they've discovered a pattern. You can take away up to 40 percent of the life-forms in a system, and the number of secondary extinctions doesn't increase significantly. "But there's a critical interval after that where things happen rapidly—a threshold effect," Roopnarine said. "The secondary extinction numbers rise dramatically."

Imagine a world like the one we live in today, with a variety of creatures in many different environments. Let's say we begin to chip away at one of those environments, like the American prairies. People clear grasslands, kill both predator and prey animals, and destroy insect pests. Still, the food web seems stable. Creatures and plants go extinct in the region, but there seems to be no ripple effect. And then, after centuries, we hit a tipping point. Forty percent of the nodes in the prairie food web have been knocked out. Suddenly, there are predators with very few prey. Catastrophic deaths among predators result: They are competing for scarce or no resources. And then a drought hits, killing the few remaining prairie grasses. Now our tiny herbivore population goes mostly extinct. We are left with few predators and virtually no prey. The already unstable food web falls apart, one death leading to another—and making the community more vulnerable to climate fluctuations.

"Don't expect the unraveling to be linear," Roopnarine warned. The deaths will be exponential. Once we hit the threshold, our food web has lost in the war of community selection. A new food web will rise up to take its place, turning the American prairie into a whole new world full of strange predators and grasses unlike any we've ever seen.

So the Permian extinction event yields a double lesson in survival. First, it offers compelling evidence that climate change caused by greenhouse gases can kill nearly every creature on the planet. Regardless of how that greenhouse scenario starts—whether it's a massive volcano or an industrial revolution—climate change can kill more effectively than a meteorite impact. Of course, atmospheric changes were only the first phase in a problem that lasted 30 million years. One could argue that food-web collapse is really what makes the Permian mass extinction a "Great Dying." The scourge started by Permian megavolcanoes echoed for millions of years, rending food web after food web until at last equilibrium was achieved in the Triassic period.

Still, there were survivors. Humans and many other mammals on Earth owe our existence to a bunch of piglike creatures with beaks who loved the sunny southern climate. That *Lystrosaurus* survived for millions of years (much longer than *Homo sapiens* has been around) proves that complex life can make it through even the most terrible disasters. These lumbering proto-mammals also left behind a few tips for what to do when we hit that wall of toxic air. By following in the lystrosaurs' footsteps, mammals dodged the next major mass extinction—even though many dinosaurs didn't.

4. WHAT REALLY HAPPENED TO THE DINOSAURS

"IT IS VERY hard to imagine what happened," the paleontologist Jan Smit said. He was describing the minutes and days following the impact of a massive meteorite, possibly 10 kilometers wide, that slammed into the Earth roughly 65 million years ago. Smit is one of the scientists who first discovered evidence for this violent event back in the 1970s. Today many of his colleagues agree that it's what caused the Cretaceous-Tertiary (K-T) mass extinction—or, as it's better known, the extinction that ended the dinosaurs.

The Cretaceous Period (145.5 Million–65.5 Million Years Ago): Meteorite Impact

Though nearly everyone is familiar with the story, Smit finds himself constantly correcting people's misconceptions about it. "It wasn't like [the movie] *Armageddon* at all," he chuckled. The Earth wasn't wrapped in fire. There were no enormous dust storms choking the life out of the soon-to-be-extinct dinosaurs. Instead, Smit said, most of the molten splash-back from the hit would have been hurled right back into space. And that's why it was so deadly.

The energy released by the meteorite slamming itself thirty meters deep into Mexico's Yucatán Peninsula was enough to punch a hole in the atmosphere. As Smit put it, "For this kind of impact, blowing away the atmosphere is a piece of cake." Tiny droplets of liquified rock and metals shot into space, quickly wreathing our stratosphere in a thick layer of extremely high clouds. The biggest problem was that the meteorite hit Earth in a particularly tender spot, geologically speaking. Beneath the Yucatán, Smit explained, "are three kilometers of limestone, dolomite, magnesium, and gypsum, and salt. Gypsum is about half sulfur. There aren't that many areas in the world that contain that much sulfur." Essentially, the meteorite vaporized a hidden cache of explosives and poisons, scattering them everywhere. Still, the mass death that swept the world afterward was not caused by acid rain or other poisons, according to Smit. Instead, it had to do with a peculiar property of vaporized sulfur: When reduced to tiny droplets in the upper atmosphere, the resulting cloud becomes highly reflective. "From space, the planet would have looked brilliantly white," Smit speculated. For at least a month, Earth became a giant reflector, and little to no sunlight could have penetrated the sulfur-laced clouds. It would have been a very extreme version of what happened after the Permian megavolcano shot sulfur into the atmosphere and cooled the planet.

Below the cloud, it would have been dark for weeks or months. Death would have come quickly to anything that drew sustenance from sunlight, including most plants. Next to die would be plant-eaters whose food sources were gone, followed by starvation among the dinosaur predators at the top of the food chain. Imagine a near-instantaneous food-web collapse, a fast-motion version of the collapses that Roopnarine described as choking off early Triassic life over millions of years. What this means is that one of the planet's most notorious disasters, complete with cinematic explosions, caused global mass extinction simply by shutting down photosynthesis.

Perhaps more than any of the other mass extinctions we've talked about so far, the K-T extinction dramatizes how mass death on Earth is tied to environmental changes. The area around the Yucatán would have been devastated after the meteorite hit. Toxic gas, fire, and extreme tidal waves would have sterilized the region around what is now called the Chicxulub crater. But even the death by darkness that followed would have been just the opening act. It would have taken centuries, and perhaps millennia, before the K-T event achieved full mass-extinction status. The dinosaurs did not die out during one long, sulfur-enhanced night. In fact, Smit underscored that the truly devastating effect of the sulfur cloud was most likely a temperature drop of 10 degrees Celsius that lasted for at least half a century, and probably a lot longer. The lush, green tropics of the Cretaceous cooled, ocean temperatures dropped, and animals who couldn't migrate found themselves trapped in hostile ecosystems. The mass extinction took out as many as 76 percent of species, including all the non-avian dinosaurs. Meanwhile, a group of mouse-sized furry creatures we know today as mammals began to thrive and grow.

The flaming-ball-of-death controversy

The K-T mass extinction is the most recent one in Earth history, and the evidence it left behind is richer than what we've got for any comparable event. As a result, scientists who study the K-T have had to confront the full complexity of life when it collapses. Not surprisingly, this has led to some of the bitterest debates in paleontology.

Smit and a UC Berkeley colleague, the geologist Walter Alvarez, endured years of doubt and ridicule when they first began speculating that the dinosaurs' demise began with a meteorite. Previously, paleontologists accounted for the mass extinction by suggesting everything from cosmic-ray bombardment to starvation. It took a while for the scientific community to accept the idea of a flaming ball from space. But Smit and Alvarez had pretty compelling evidence. Working on opposite sides of the globe—Smit in Spain, and Alvarez in the Americas—the two researchers uncovered physical remains of the impact and overturned the previous theories about why dinosaurs suddenly went extinct after ruling the planet for over 100 millennia. Alvarez worked with his Nobel Prize–winning physicist father, Luis Alvarez, and the two published a history-making paper in 1980 showing that rock layers at the K-T boundary contained a high concentration of iridium, a metal found almost exclusively in space. Meanwhile, Smit had discovered "spherules," tiny balls of rock that had been heated up and then cooled down quickly, in the same layer all over the world. Smit published a paper about the spherules the same year Alvarez published his about what's come to be known as the "iridium anomaly." The one-two punch of these papers—chronicling metals from space and the remains of superheated rock scattered across the planet—suggested an event whose magnitude could easily have accounted for global mass death.

But the flaming-ball controversy is still far from over. In the late 1980s, Princeton geologist Gerta Keller began publishing papers questioning whether the meteorite impact actually had a global effect after all. She claimed she had a better explanation: megavolcanoes in India. And she spent the next two decades gathering evidence to prove her hypothesis, despite widespread scorn from the scientific community. UC Berkeley paleontologist Charles Marshall said that "nobody in the scientific community takes [Keller] seriously," and Smit told the BBC that her ideas "are barely scientific." Like Smit and Alvarez before her, Keller cheerfully met doubt with documentation.

When I spoke to Keller, she had recently returned from India, where she'd made a series of incredible discoveries. She and a group of local scientists managed to get samples three kilometers deep underground in a region called the Deccan Plateau, long known to be the site of an ancient megavolcano. The area has been off limits to scientists ever since India's Oil and Natural Gas Corporation started drilling there. An outspoken person who clearly loves a good scientific fight, Keller put her considerable powers of persuasion into a campaign to gain access to what she suspected might reveal the truth about how the dinosaurs died. She finally got her wish and, joined by a group of Indian scientists, she found more than she'd ever hoped.

Using special tools that produce "cores," cylindrical rock samples pulled up from deep underground using cannulated drills, Keller and her colleagues discovered that the Deccan Plateau was the result of at least four major eruptions following closely on each other. One eruption was so enormous that the team found a single uninterrupted lava flow stretching 1,500 kilometers from the volcanic vent all the way to the sea.

But the most valuable discovery was the layers of sediment in between each lava flow–in those sediments, Keller and her colleagues found fossils that helped date the volcanic eruptions. Based on the evidence so far, it appears that the Deccan supervolcano began spewing lava and toxic gas about 67.4 million years ago–about 1.6 million years before the K-T boundary. The timing was right. And so were the deadly patterns of extinction she observed in those layers of sediment. After each lava flow, fewer and fewer animals were recovering from the devastation. "By the time the fourth flow came, nothing was left," she said.

Keller believes that the flows may have come so rapidly that life nearby had no chance to recover–and that the toxins and carbon released by the explosions wound up killing off creatures across the globe with environmental changes similar to those at the end of the Permian. A runaway greenhouse effect, combined with acid rain and ocean dead zones, would have made the planet unlivable for the majority of its inhabitants. "That's the likely killing mechanism," Keller concluded matter-of-factly.

Who is right? It's entirely possible that both Smit and Alvarez on one side and Keller on the other have identified causes of the K-T mass extinction. There are other theories, too. A fungal spike in the fossil record during the mass extinction has led at least one scientist to suggest that the dinosaurs died of fungal infections like the ones that are causing extinctions among amphibians and bats today. When the evidence in the geological record is relatively fresh, it becomes obvious that most mass extinctions on Earth have multiple causes. And as Keller's work suggests, evidence gathered outside Europe and the Americas can offer a new perspective on old theories. We know that the bodies start piling up when environments change, but many events all over the world may have set those changes in motion.

The Late Triassic Period (220 Million–200 Million Years Ago): The Beginning of Our World

One of the difficulties in sorting out what happened to the dinosaurs has nothing to do with geological evidence and everything to do with human culture. Dinosaurs have been so widely misrepresented in pop culture about the prehistoric world that it's hard for us to step back and appreciate this diverse array of creatures for what they actually were, and how they really died out.

To get the real story, we'll return to the chaotic Triassic period that followed the Great Dying, when many species evolved and died out rapidly. Late in the Triassic, about 220 million years ago, dinosaurs began to evolve. At this time, they were, as Brown University geologist Jessica Whiteside put it, "about the size of German shepherds and not very diverse." Their main competitors were the crurotarsans, those fierce carnivores that eventually evolved into crocodiles and alligators. How did a relatively small group of mini-dinos prevail against these toothy, occasionally armored beasts? "If you were in the Triassic, you would bet on crurotarsans," Whiteside said. But surprisingly, most crurotarsans did not survive the mass extinction that ended the Triassic, leaving the dinosaurs to take over lands once dominated by their mega-gator counterparts.

Whiteside attributes this bizarre turn of events to one of the most stupendous underwater volcanoes in Earth history. Known as the Central Atlantic magmatic province (CAMP), the eruption started about 200 million years ago in a narrow body of water separating the eastern Americas from West Africa. (At that time, the two continents were still joined into the Pangaea supercontinent.) The lava flow from CAMP was tremendous. It forced the continental plates so far apart that an entire ocean grew between once-connected regions known today as Canada and Morocco.

If this eruption could create one of Earth's biggest oceans, just imagine the high volumes of carbon, methane, and sulfur it was pumping into the water and the atmosphere. A superhot greenhouse gripped the planet as the Triassic wound to a murky close. Whiteside ticked off the deaths that followed: As the temperatures climbed higher, the world-spanning tropical forests of the Triassic dried out and succumbed to enormous wildfires. The burned remains of forests slipped into the oceans along with carbon-rich soil. The oceans became acidic, which led to anoxia and die-offs there. Coral reefs were the first to go, and their deaths set off a cascading effect where anything that fed higher in the food chain died too. It was the perfect storm for destroying food webs, starting in the oceans and creeping up onto a warming land whose trees were being eaten by fire. Once again, climate change was killing the world.

The rise (and fall and rise) of the dinosaurs' world

There are many well-preserved plant fossils from this era, so it's possible to visualize how the extreme greenhouse conditions changed the environment between the Triassic and the Jurassic. Jennifer McElwain, a paleobotanist at University College Dublin, has spent years studying this transition in Greenland, excavating everything from leaves and flowers to microscopic bits of pollen, to reconstruct the world where dinosaurs ultimately triumphed. Today, coastal Greenland is hard tundra that's too cold for trees, but in the late Triassic and early Jurassic, it was full of lush vegetation. McElwain called it "a cross between New Zealand conifers and the Florida Everglades." It was a world of "broad, meandering rivers" and "big, wide floodplains" bordering forests full of towering trees and stubby, thick-trunked plants called cycadeoids with palmlike fronds bursting from their tops. And then came CAMP, with its carbon emissions and rising global temperatures.

Tens of thousands of years of greenhouse conditions led to fire after fire. Ultimately, McElwain believes, the environment of diverse trees, shady forests, and thick vegetation was reduced to swamps full of ferns. "There would have been ferns as far as the eye can see, with hardly any trees, and lots of fire," McElwain said. There was no complex, multi-tiered canopy in the forests, so the landscape would have been much brighter. But within another 100,000 years, the region went back to being conifer-dominated. What emerges from this fast-motion vision of ancient forests rising, burning, and rising again is something approaching the truth of where the dinosaurs began. They were among the only survivors of radical environmental changes that drove their competitors to extinction. Most crurotarsans were extinguished in the burned threads of food webs, but those small, early dinosaurs were able to spread out and adapt to the new environments and continents.

When forests at last returned to the land, dinosaurs evolved to be much larger. They diversified into armored herbivores like triceratops and plate-backed stegosaurus, sneaky scavengers, and predators like the 40-foot-long *T. rex* that we've seen in movies from the 1933 version of *King Kong* to *Jurassic Park*. Dinosaurs were as diverse as mammals are today, and their behavior probably varied a great deal from species to species. Many of them walked on two legs, with body postures similar to birds—their heads would have been thrust far forward, their spines nearly horizontal, and their tails held out stiffly behind them rather than dragging on the ground. Indeed, most paleontologists today accept that birds evolved from therapods, a group of bipedal, feathered dinosaurs that included the infamous *T. rex*. If you ever want to imagine what it would be like to face down a dinosaur, imagine a hulking, 40-foot-long crow whose beak has become a toothy mouth.

Recent evidence suggests that many dinosaurs weren't feathered in quite the way birds are today. Most had dark gray or reddish proto-feathers (often called dinofuzz) that looked something like spiny down. Indeed, dinosaurs may have had proto-feathers for millions of years before birds evolved the ability to fly. Also like their bird relatives, many dinosaurs made nests and laid eggs. Though it's hard to piece together how these different Cretaceous animals might have behaved, some paleontologists theorize that they may have been social, like birds, forming flocks and possibly mating for life.

What we do know is that when the catastrophes of the Cretaceous period hit, dinosaurs were in a position similar to their old rivals, the mega-gator crurotarsans. A lot of dinosaurs had evolved into specialists, and were therefore deeply connected to food webs that were all too easy to unravel with a few shifts in global temperature. This time, a group of mouse-like, furry animals called mammals—the descendants of the Permian survivor *Lystrosaurus*—were the survivors.

The Earth these mammals began to colonize with their strange paws and non-feathered faces had come into being through extremely complex events, whose true impact can only be measured in tens of millions of years. Environments had died and been reborn from the effects of liquid rock deep in the Earth and flaming balls from space; the mixture of gases in the atmosphere had been altered by microbes, mountains, and plants; temperatures had fluctuated between extreme icehouses and greenhouses; and the very shape of the planet's landmasses and oceans had transformed quite radically dozens of times. If there had been a paleogeologist among the last of the dinosaurs, she could hardly have pinned the blame for her peers' demise on any single factor. The entire ambiguous history of the planet would have had to stand trial for murdering brachiosaurus and letting a bunch of little monkeys take over.

The dinosaurs who survived

Adding to our paleogeologist dinosaur's conundrum would be another issue, which is that the dinosaurs didn't entirely die out. An evolutionary offshoot of therapods—birds—survived into the present to become one of the most successful animal classes on the planet. They are highly diverse, exhibit a wide variety of social behaviors, and undertake some of the most incredible migratory journeys of any creatures we know. Of course their dinosaur forebears are extinct, much the way humans' forebears are. But the dinosaur evolutionary line appears to have continued unbroken.

Perhaps the single most common misconception about dinosaurs among humans is that these creatures and their world have disappeared. Like mammals, dinosaurs are survivors. But their children, who flash through the skies and leave us in awe, are so different from their ancestors that we find it hard to draw a connection between them. What we should ponder, as we move from geological history into the world where human evolution takes place, is whether our understanding of survival is as clouded as our understanding of dinosaurs.

The dinosaurs survived two mass extinctions, but the crows who like to hang out in the tree next to my house are nothing like those dog-sized dinosaurs who beat out the crurotarsans. In fact, it's not entirely accurate to say the crurotarsans have been pushed off the environmental stage either. Are we not witnessing a strange tableau of survival whenever a bird alights on the head of a crocodile, bringing together the evolutionary offspring of Triassic and Jurassic? Instead of saying the dinosaurs died out, it might be more accurate to say that dinosaurs changed.

Can humans possibly expect to remain unchanged as we face the next mass extinction? History suggests that it's unlikely. But if survival means that our species will evolve into creatures like ourselves, but with new abilities–like, say, flight–that's not so bad. Some would even call it an improvement. Survival may be far weirder, and better, than we ever imagined.

5. IS A MASS EXTINCTION GOING ON RIGHT NOW?

IN THE OREGON high desert, a dark, fissured ridge bulges above the broad expanse of dusty ground and blue-green scrub. Observed from a hundred yards away, it looks like nothing more than a rocky outcropping. But viewed from the perspective of thousands of years in the past, it's a landmark of incredible import. Hiding beneath the brows of this ridge are the wide, low entrances to the Paisley Caves, generous shelters that humans used as a rest stop for thousands of years. Over the past decade, the University of Oregon archaeologist Dennis Jenkins has led excavations in these caves that unearthed evidence of human habitation dating back over 14,000 years. That makes this one of the oldest known human campsites in the Americas.

The Paisley Caves mark a significant moment in human history. Many scientists identify the people who first came here as harbingers of a new mass extinction, authored by *Homo sapiens,* that's started to accelerate during the past three centuries. The remains that litter the Paisley Caves include bones from some of the first animals that humans may have driven to extinction: mastodons and mammoths (often dubbed "megafauna" or "megamammals"), as well as American horses and camels. For millions of years, such creatures had dominated the continents' vast plains and forests; soon, humans would claim these environments as their own. When humans were building cooking fires in the Paisley Caves, our species was on the verge of becoming populous enough to push other creatures out of their native habitats. Over the next few millennia, the population exploded. Humans invaded new habitats, pushing the bigger mammals out. We also killed these animals outright. Megafauna were a big part of the Neolithic diet. We find evidence of this in charred, gnawed bones that early settlers left behind, as well as in cave paintings that depict mammoth hunts.

But at the time humans first spent the night in the Paisley Caves, megafauna roamed the Oregon mountains and the environment was much lusher and wetter than today. The Americas, and indeed the planet, had not yet been significantly transformed by human incursions. Jenkins described what he imagined was a typical view from the caves 14,000 years ago. The vast, dusty plains around the cave mouths today would have sparkled with water, where camels and mastodons came to drink. Even then, the caves wouldn't have made an ideal village—they were too far from the water's edge. "People would come to these caves periodically, but it wasn't home," Jenkins explained. It was a Neolithic rest area between two swampy regions that were packed with food and water. Enough people traveled between the two areas for thousands of years that their overnight stopovers left layers of detritus from campfires, tools, and waste. These caves are evidence that humans were trekking all over the place 14,000 years ago. While their brethren back in the lands that later became Syria and Turkey were erecting some of the first temples and proto-cities, these people were the first explorers in rich, uncharted land.

As humans spread across the American continents, starting from boats and coastal outposts along the Pacific Rim and working their way inland, they pushed the native wildlife out. By 10,000 years ago, most of the American megafauna were dead.

UC Berkeley biologist Anthony Barnosky has been at the forefront of research into megafauna extinctions and their relationship to a possible sixth mass extinction today. A careful scholar, he's also an activist who is as at home talking to environmentalists on Twitter as he is in the pages of the prestigious journal *Nature*. He believes that the signs of mass extinction are all around us, and have been for millennia– which is really the only scale on which we can measure mass extinctions anyway. This sixth extinction began with the megafauna, which Barnosky believes weren't simply victims of human hunting and expansion–there was also climate change from a minor ice age called the Older Dryas that would have decimated the beasts' favored grazing grounds. If we are in a mass extinction, he concluded, it was kicked off by a "synergy of climate change and humans . . . the combination was evidently very bad." Over the past few centuries, industrialization and human population explosions have changed the landscape further. There is ample evidence that we've driven dozens of species to extinction. But can we really call this a mass extinction, comparable to the end of the Cretaceous or Permian?

In a widely read article published in the March 3, 2011, issue of *Nature*, "Has the Earth's Sixth Mass Extinction Already Arrived?" Barnosky and many of his colleagues (including the statistician and paleontologist Charles Marshall) argued that we can. In it, they explained that the extinction rates on Earth today are far above the background rate. If today's endangered species all go extinct, our planet will be in the grip of a mass extinction within the next 200 years. Within 1,000 years, Earth might be a world as changed as it was after each of the previous mass extinctions we've discussed. The problem, Barnosky admits, is trying to pin down whether we're in the middle of a mass extinction when such events are usually measured on an extremely long timescale.

"I think we're on a leading edge," Barnosky told me. "My take on it is that we're actually not far into it. A true mass extinction is losing seventy-five percent of species that are recorded. We've lost maybe one or two percent of those we can count. So everything we want to save is still out there." Still, he cautioned, the big problem is not our world right now but the world we're heading toward over the next century. Looking at the data, he and his colleagues believe that extinction rates for mammals are far above the typical extinction numbers we'd expect for a background rate. "It's happening too fast," he sighed. "We're somewhere between three and twelve times too high." Given that humans are only likely to expand our territories further into those of endangered animals, he expects these numbers to grow. And when you add in all the carbon we're pumping into the atmosphere, it's possible that we're re-creating the conditions that led to previous mass extinctions.

Peter Ward, a geologist at the University of Washington, who has written about mass extinctions in several books, including his influential work *The Medea Hypothesis,* believes carbon emissions mean that environmental change is almost inevitable. "We're going back to the Miocene," he said, then laughed darkly. The Miocene, a geological age that ended roughly 5.3 million years ago, was the last time that the planet had no Arctic ice cap. It was a period of intense heat when greenhouse conditions reigned and our hominid ancestors had not yet evolved. Though many animals might thrive in the Miocene climate, humans wouldn't. We are the products of a cold Earth, just like many of our mammal brethren. "We need to keep those ice caps," Ward said.

The question for scientists like Barnosky and Ward is whether somebody living millions of years from now could look back on our own geological period, the Quaternary, and say that it ended with the sixth mass extinction event on our planet. If so, that would put humans in a class with cyanobacteria as the only life-forms that ever single-handedly brought on an environment-changing event with widespread deadly effects. However, as we've seen from looking at previous mass extinctions, it's impossible to pin the blame for such an enormous event on just a single catastrophe—or a single species' meddling.

If we are in the early days of a mass extinction, the main thing that sets it apart from the five previous ones is the presence of a species that has the ability to stop it. We are tenacious survivors, incredible inventors, and we've demonstrated an ability to plan for the future collectively—even, sometimes, for the good of all rather than the good of the few. One of our most powerful skills in making those plans is our knowledge of history. Not only have we kept records of human history for thousands of years, but we've also developed scientific methods of discovering what happened to the planet before we evolved. The geological history we've just shot through at top speed is full of information about the kinds of life-eradicating dangers that Earth has confronted over and over. By remembering this history, we can make informed decisions about what to do next in order to ensure our survival as a species.

In the next section of the book, we'll explore how humans have already made it through tens of thousands of years of environmental catastrophe, disease, and famine. With each blow to our species, we've crafted better and better methods of surviving.

PART II WE ALMOST DIDN'T MAKE IT

6. THE AFRICAN BOTTLENECK

MOST OF US are familiar with the basic outlines of the human evolutionary story. Our distant ancestors were a group of apelike creatures who started walking upright millions of years ago in Africa, eventually developing bigger brains and scattering throughout the world to become the humans of today. But there's another story that has received less attention. Advances in genetics have given us a sharper understanding of what happened between the "walking upright" and the "buying the latest tablet computer" chapters of the tale.

Written into our genomes is the signature left behind by an event when the early human population dwindled to such a small size that our ancient ancestors living in Africa may have come close to extinction. Population geneticists call events like these bottlenecks. They're periods when the diversity of a species becomes so constrained that evidence of genetic culling is obvious even thousands of generations later. Sometimes the shrinking of a population is the result of mass deaths, and indeed, there is evidence that humans may have been fleeing a natural disaster when we walked out of Africa roughly 70 thousand years ago. But our species probably experienced multiple genetic bottlenecks beginning as far back as 2 million years. And those earlier bottlenecks were caused by a force far more powerful than mass death: the process of evolution itself.

In fact, the African bottlenecks are an example of the paradoxical nature of human survival. They provide evidence that humans nearly died out many times, but also tell a story about how we evolved to survive in places very far away from our evolutionary home in Africa.

The Fundamental Mystery of Human Evolution

Given our enormous, globe-spanning population size, humans have remarkably low genetic diversity—much lower than other mammal species. All 6 billion of us are descended from a group of people who numbered in the mere tens of thousands. When population geneticists describe this peculiar situation, they talk about the difference between humanity's actual population size and our "effective population size." An effective population size is a subgroup of the actual population that reasonably represents the genetic diversity of the whole. Put another way, humanity is like a giant dance party full of billions of diverse people. But population geneticists, elite party animals that they are, have managed to find the one ideal VIP area that contains a small group of people who very roughly capture the diversity of the party as a whole. In theory, that room contains the party's effective population size. If they all started randomly having sex with each other, their children might loosely reproduce the diversity and genetic drift of our actual, billions-strong population.

The weird part is that compared with our actual population size, the human effective population in that VIP area is very low. In fact, today's human effective population size is estimated at about 10,000 people. As a point of comparison, the common house mouse is estimated to have an effective population size of 160,000. How could there be so many of us, and so little genetic diversity?

This is one of the fundamental mysteries of human evolution, and is the subject of great debate among scientists. There are a few compelling theories, which we'll discuss shortly, but there is one point that nearly all evolutionary biologists will agree on. We are descended from a group of proto-humans who were fairly diverse 2 million years ago, but whose diversity crashed and passed through a bottleneck while *Homo sapiens* evolved. That crash limited our gene pool, creating the small effective population size we have today. Does some kind of terrible disaster lurk in the human past? An event that could have winnowed our population down to a small group of survivors, who became our ancestors? That's definitely one possibility. Evolutionary biologist Richard Dawkins has popularized the idea that the population crash came in the wake of the Toba catastrophe, a supervolcano that rocked Indonesia 80,000 years ago. It's possible this enormous blast cooled the African climate for many years, destroying local food sources and starving everybody to death before sending fearful bands of *Homo sapiens* running out of Africa.

But, as John Hawks, an anthropologist at the University of Wisconsin, Madison, put it to me, a careful examination of the genetic evidence doesn't reveal anything as dramatic as a single megavolcanic wipeout. Instead of some Hollywood special-effects extravaganza, human history was more like a perilous immigration story. To understand how immigration can turn a vast population into a tiny one, we need to travel back a few million years to the place and time where we evolved.

The Human Diaspora

Humanity's first great revolution, according to the anthropologist Ian Tattersall of the American Museum of Natural History, was when it learned to walk upright, more than 5 million years ago. At the time, we were part of a hominin group called *Australopithecus* that shared a very recent common ancestor with apes. Australopithecines hailed from the temperate, lush East African coast. They were short–about the size of an eight-year-old child–and covered in a light layer of fur. They may have started walking on their hind legs because it helped them hunt and find the fruits that dominated their diets. Whatever the reason, walking upright was unique to *Australopithecus*. Her fellow primates continued to prefer a four-legged gait, as they do today.

Over the next few million years, *Australopithecus* walked from the tip of what is now South Africa all the way up to where Chad and Sudan are today. Our ancestors also grew larger skulls, anticipating a trend that

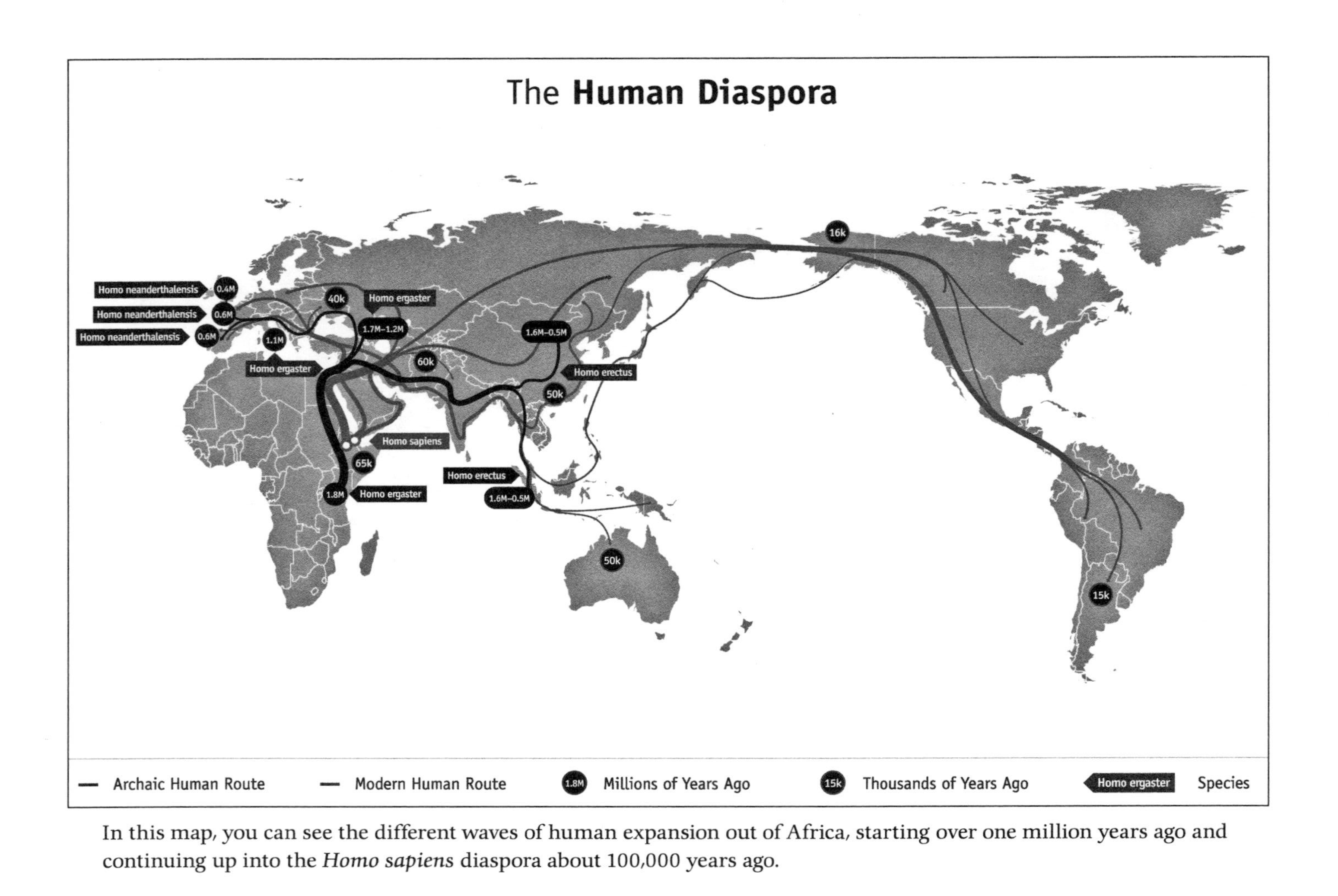

In this map, you can see the different waves of human expansion out of Africa, starting over one million years ago and continuing up into the *Homo sapiens* diaspora about 100,000 years ago.

has continued throughout human evolution. By about 2 million years ago, *Australopithecus* was evolving into a very human-looking hominin called *Homo ergaster* (sometimes called *Homo erectus*). Similar in height to humans today, a couple of *H. ergaster* individuals could put on jeans and T-shirts and blend in fairly well on a typical city street—as long as they wore hats to hide their slightly prominent brows and sloping foreheads. Another thing that would make our *H. ergasters* feel perfectly comfortable loping down Market Street is the way so many in the crowd around them would be clutching small, hand-sized tools. Our tools may contain microchips whose components are the products of advanced chemical processing, but the typical smartphone's size and heft are comparable to the carefully crafted hand axes that anthropologists have identified as a key component of *H. ergaster*'s tool kit. *H. ergaster* wouldn't need anyone to explain the meat slowly cooking over low flames in kebab stands, either: There's evidence that their species had mastered fire 1.5 million years ago.

There are many ways to tell the story of what happened to *H. ergaster* and her children, who eventually built those smart phones and invented the tasty perfection that is a kebab. *H. ergaster* was one of many bipedal, tool-using hominids roaming southern and eastern Africa who had evolved from *Australopithecus*. The fossil record from this time is fairly sparse, so we can't be sure how many groups there were, what kinds of relationships they formed with each other, or even (in some cases) which ones evolved into what. But each group had its own unique collection of genes, some of which still survive today in *Homo sapiens*. And those are the groups whose paths we're going to follow.

This path is both a physical and a genetic one. A visitor to the American Museum of Natural History in New York can track its progress in fossils. Glass-enclosed panoramas offer glimpses of what we know about how *H. ergaster* and her progeny created hand axes by striking one stone against another until enough pieces had flaked off that only a sharp blade was left. Reconstructed early human skeletons stand near sparse fossils and tools, a reminder that our ideas about these people come, literally, from mere fragments of their bodies and cultures. Ian Tattersall has spent most of his career poring over those fragments, trying to reconstruct the tangled root structure of humanity's evolutionary tree.

One thing we know for sure is that early humans were wanderers. Not only did they spread across Africa, but they actually crossed out of it many times, starting about 2 million years ago. Anthropologists can track the journeys taken by *H. ergaster* and her progeny by tracing the likely paths between what remains of these peoples' campsites and villages, often identifying the group who lived there based on the kinds of tools they used.

Tattersall believes there were at least three major radiations, or population dispersals, out of Africa. Despite the popularity of Dawkins's Toba volcano theory, Tattersall believes there was "no environmental reason" for these immigrations. Instead, they were all spurred by evolutionary developments that allowed humans to master their environments. "The first radiation seems to have coincided with a change in body structure," he mused. Members of *H. ergaster* had a more modern skeletal structure featuring longer legs than their hominid cohorts, which meant they could walk quickly and efficiently over a variety of terrains. Tattersall explained that there were environmental changes in Africa during this time, but not enough to suggest that humans fled environmental destruction to greener pastures. Instead they were simply well suited to explore "unfamiliar environments, ones very unlike their ancestral environments," he said. *H. ergaster*'s rolling gait was an adaptation that allowed the species to continue adapting, by spreading into new lands where other hominids literally could not tread.

As early humans walked into new regions, they separated into different, smaller bands. Each of these bands continued to evolve in ways that suited the environments where they eventually settled. We're going to focus on four major players in this evolutionary family drama: our early ancestor *H. ergaster* and three siblings she spawned—*Homo erectus, Homo neanderthalensis,* and *Homo sapiens.*

H. erectus was likely the evolutionary product of that first exodus out of Africa that Tattersall described. About 1.8 million years ago, *H. erectus* crossed out of Africa through what is today Egypt and spread from there all the way across Asia. These hominins soon found themselves in a very different environment from their siblings back in Africa; the winds were cold and snowy, and the steppes were full of completely unfamiliar wildlife. Over the millennia, *H. erectus*'s skull shape changed and so did her tool sets. We can actually track how our ancestors' tools changed more easily than how their bodies did because stone preserves better than bone. Scientists have reconstructed the spread of *H. erectus* by unearthing caches of tools whose shapes are quite distinct from what other groups used. From what we can piece together, it seems that *H. erectus* founded cultures and communities that lasted for hundreds of thousands of years, and spread throughout China and down into Java.

Over the next million years, other groups of humans followed in *H. erectus*'s footsteps, walking through Egypt to take their siblings' route out of Africa. But as the Stanford paleoanthropologist Richard Klein told me, these journeys probably weren't distinct waves of migration. Walking in small groups, these humans were slowly expanding the boundaries of the hominin neighborhood.

Fossil remains in Europe suggest that about 500,000 to 600,000 years ago, some of *H. ergaster*'s progeny, on emerging from Africa, decided to go left instead of right, wandering into the western and central parts of the Eurasian continent. These Europeans evolved into *H. neanderthalensis*. They often set up homes in generously sized cave systems, and there's evidence that some groups lived for dozens of generations in the same caves, scattered throughout Italy, Spain, England, Russia, and Slovenia, among other countries. Neanderthals evolved a thicker brow and more barrel-chested body to cope with the colder climate. We'll talk more about them in the next chapter.

Back in Africa, *H. ergaster* was busy, too, establishing home bases all over the coasts of the continent, reaching from southern Africa all the way up to regions that are today Algeria and Morocco. By 200,000 years ago, *H. ergaster*'s skeletal shape was indistinguishable from that of modern humans. A species we would recognize as *H. sapiens* had emerged. And that's when human beings made their next evolutionary leap—one that perfectly complemented our ability to walk upright into new domains.

How We Evolved to Tell Stories

"When *Homo sapiens* came along there was something totally radical about it," Tattersall enthused. "For a hundred thousand years, *Homo sapiens* behaved in basically the same ways its ancestors had. But suddenly something happened that started a different pattern." Put simply, humans started to use the giant brains they'd evolved to fit inside their gradually enlarging craniums. What changed? Tattersall said there are no easy answers, but evolution often works in jumps and starts like that. For example, birds evolved feathers millions of years before they started flying, and animals had limbs long before they started walking. "We had a big brain with symbolic potential before we used it for symbolic thought," he concluded. In what anthropologists call a cultural explosion over the past 100,000 years, humans developed complex symbolic communication, from language and art to fashion and complex tools. Instead of looking at the world as a place to avoid danger and score food, humans disassembled it into mental symbols that allowed us to imagine new worlds, or new versions of the world we lived in.

Humans' new facility with symbols allowed us to learn about the world around us from other humans rather than starting from scratch with direct observations each time we went to a new place. Like walking, symbolic thought is an adaptation that leads to more adaptations. Modern humans could venture into new territory, discover its resources and perils, then tell other bands of humans about it. They might even pass along designs for tools that helped us gain access to foods specific to a certain area, like crushers for nuts or scoops for tubers. Aided by our new capacity for imagination, those bands of humans could familiarize themselves with alien regions before ever visiting them. For the first time in history, people could figure out how to adapt to a place before arriving there—just by hearing stories from their comrades. Symbolic thought is what allowed us to thrive in environments far from warm, coastal Africa, where we began. It was the perfect evolutionary development for a species whose body propelled us easily into new places. Indeed, one might argue that the farther we wandered, the more we evolved our skills as storytellers.

Let's go back, for a moment, to that first radiation out of Africa, nearly 2 million years ago when *H. ergaster*, with her small but effective tool kit, crossed into the Sinai Peninsula. At roughly the same time, we find evidence of humanity's first genetic bottleneck. And yet, as Tattersall and many others have pointed out, there is no evidence of a giant disaster thinning the population, leaving the survivors to flee across the Middle East and Asia. The bottleneck is clearly a sign of a population crash, but what caused it? As I said earlier, the effective population size for *H. sapiens* is estimated at roughly 10,000 individuals; but the University of Utah geneticist Chad Huff recently argued that soon after *H. ergaster* left, our effective population size was about 18,500. It's likely this bottleneck is actually a record of human groups growing smaller as they thinned out across the Eurasian continent, meeting adversity every step of the way. At the same time, according to anthropologist John Hawks, the bottleneck is a mark of evolutionary changes that could only happen to a population that was always on the move.

It started with that first trek out of Africa, which split humanity into several groups. As Hawks explained in a paper he published with colleagues in 2000, one cause for a genetic bottleneck can be speciation, or the process of one species splitting into two or more genetically distinct groups. We've already touched on how *H. ergaster* evolved into at least three sibling groups, but that's a dramatic oversimplification. For example, *H. ergaster* likely evolved into a group called *Homo heidelbergensis* in Africa, which then speciated into *H. sapiens* and another group that speciated into Neanderthals and their close relatives the Denisovans later on. There are many complexities in the lineage of *H. erectus*, too, especially once the group reached Asia. Evolution is a messy process, with many byways and dead ends. By the time *H. ergaster* reached the Sinai, the group would have undergone at least one speciation event–the one that led to early *H. erectus*. That means only a subset of *H. ergaster* genes survived in *H. erectus*, and a subset of its genes survived in the *H. ergaster* groups who stayed behind. If these groups remained small, and there's ample reason to believe that they did, you now have two isolated gene pools that are less diverse than the original one. That's how speciation creates a genetic bottleneck.

But even without speciation events, humans' habit of walking all over the place would have caused a bottleneck. The very act of wandering far from home, into many dangers, can shrink both the population and the gene pool over the course of generations. Population geneticists call this process the founder effect. To see how the founder effect works, let's follow a band of *H. erectus* passing through lands edging the Mediterranean Sea and finding its way into India. Remember, this isn't one long trek. Maybe the coast of today's state of Gujarat appeals to a few members of *H. erectus,* and so a band decides to settle down for a while in that region. This settlement is called a founder group, and it has less diversity than the group it came from simply because it has fewer members. In the next generation, a new group splits off from the Gujaratis and heads south along the coast. Generally we assume that each time a group left for untouched lands, it left a group behind. So each new group becomes a founder population in its own right, and has less genetic diversity than the group back in Gujarat—even if you factor in some intermarriage between different founder groups. Multiple founder events in a row would have had the odd effect of increasing humanity's population while decreasing human genetic diversity. Now, consider the fact that our *H. erectus* explorers in India are a microcosm of the way all humans spread across the Earth. After hundreds of generations of wandering, humans managed to increase their populations gradually while retaining the low diversity caused by genetic bottlenecks.

Back in Africa, early humans were also speciating and wandering, forming new bands, each of whose genetic diversity was lower than the last. But when a small band of hominins called *H. sapiens* evolved, about 200,000 years ago, something strange happened. Tattersall believes that humans underwent some kind of genetic change that spurred a cultural shift. Suddenly, between 100,000 and 50,000 years ago, the fossil record is full of sculpture, shell jewelry, complex tools made from multiple kinds of material, ochre-and-carbon cave paintings, and elaborate burial sites. Possibly, as Randall White, an anthropologist at New York University, suggests in his book *Prehistoric Art,* humans were using jewelry and clothing to proclaim allegiance with particular groups. *H. sapiens* wasn't just interacting with the world. They were using symbols to mediate their relationship with it. But why the sudden shift from a hominin with the capacity for cultural expression to a hominin who actively created culture?

It could be that one small group of *H. sapiens* developed a genetic mutation that led to experiments with cultural expression. Then, the capacity to do it spread via mating between groups because storytelling and symbolic thought were invaluable survival skills for a species that regularly encountered unfamiliar environments. Using language and stories, one group could explain to another how to hunt the local animals and which plants were safe to eat. Armed with this information, humans could conquer territory more quickly. Any group that could do this would have a higher chance of surviving relocation time and again. The more those groups survived, the more able they were to pass along any genetic predisposition for symbolic communication.

Perhaps *H. sapiens'* knack for symbolic culture was also a result of sexual selection, in which certain genes spread because their bearers are more attractive to the opposite sex. Put simply, these attractive people get laid more often, and therefore have more chances to spread their genes to the next generation. In his book *The Mating Mind,* evolutionary psychologist Geoffrey Miller argues that among ancient humans, the most attractive people were good with language and tools. The result would be a population in which sexual selection created successively more symbol-oriented people. Two anthropologists, Gregory Cochran and Henry Harpending, amplify this point. They argue that some of the genes that spread like wildfire through the human population over the past 50,000 years are associated with cranial capacity–brain size–and language ability. "Life is a breeding experiment," Cochran and Harpending write in their book *The 10,000 Year Explosion.*

Our capacity for symbolism evolved quickly, partly because our mating choices would have been shaped by our needs as creatures who evolved to survive by founding new communities. Over the past million years, humans bred themselves to be the ultimate survivors, capable of both exploring the world and adapting to it by sharing stories about what we found there.

How Can We Possibly Know All This?

A lot of the evidence we have for the routes that humans took out of Africa comes from objects and places you can see with your own eyes. Paleontologists have found our ancestors' ancient bones, as well as their tools. To figure out the ages of these tools and skeletons, we use the same kinds of dating techniques that geologists use to discover the history of rocks. In fact, when an anthropologist talks about "dating the age of fossils," he or she isn't actually talking about the bones themselves–to date old bones, anthropologists carefully excavate them and take samples of the rock surrounding them. Then they pin a date on those rocks, under the assumption that the bones come from roughly the same era as the rocks or sand that covered them up. Basically, we date fossils by association, which is why you'll often hear scientists suggesting that a particular fossil might be between 100,000 and 80,000 years old. Though we can't pin an exact month or year on each fossil discovery, we do have ample evidence that certain humans like *H. ergaster* came before other humans like *H. erectus* in evolutionary and geological time.

Over the past decade, however, the study of ancient bones has been revolutionized by new technologies for sequencing genomes, including DNA extracted from the fossils of Neanderthals and other hominins who lived in the past 50,000 years (sadly, we don't have the ability to sequence DNA from *Australopithecus* or *H. ergaster* bones–their DNA is too decayed). At the Max Planck Institute in Leipzig, Germany, an evolutionary geneticist named Svante Pääbo and his team have developed technology to extract nearly intact genomes from Neanderthal bones. First they grind the bones to dust and chemically amplify whatever DNA molecules they can find, then analyze this genetic material using the same kinds of sequencers that decode the DNA of living creatures today. We'll deal with the Neanderthal genome more in the next chapter, but suffice it to say that we have pretty solid evidence about the genetic relationships between *H. sapiens* and its sibling species *H. neanderthalensis*.

A lot of the evidence for humans' low genetic diversity has been made possible by DNA-reading technologies developed since the first human genome was sequenced, in the early 1990s. Though that first human genome took over a decade to sequence, we now have machines capable of reading the entire set of letters making up one genome in just a few hours. As a result, population geneticists are accumulating a diverse sampling of sequenced human genomes, from people all over the world. Many of these genomes are collected into data sets that scientists can feed into software that does everything from make very simple comparisons between two genomes (literally analyzing the similarities and differences between one long string of letters and another), to extremely complex simulations of how these genomes might have evolved over time.

One of the first pieces of genetic evidence for the serial-founder theory emerged when scientists had collected DNA sequences from enough people that we could start to analyze genetic diversity in different regions all over the world. Geneticists discovered a telltale pattern: People born in Africa and India tend to have much greater genetic diversity than people born elsewhere. This is precisely the kind of pattern you'd expect to see in a world population that grew out of founder groups originating in Africa. Remember, each successive founder group has less and less genetic diversity. So people descended from groups that stayed in Africa or India are from early founder groups. People in Europe, Australia, Asia, and the Americas were the result of hundreds of generations of founder effects—so we'd expect them to have less genetic diversity. When you add this genetic evidence to the physical evidence from fossils and tools left behind by people leaving Africa, you wind up with a fairly solid theory that founder effects created our genetic bottleneck.

An Eruption That Launched Humanity

Though it's likely that the genetic bottlenecks we observe in the human population were caused mostly by founder effects and sexual selection, there is some evidence that the final human radiation out of Africa was precipitated by a catastrophe. Ancient humans had been crossing the Sinai out of Africa and into the rest of the world for over a million years, but roughly 80,000 years ago there was an extremely large migration that changed the world and every human on it. *H. sapiens*, a human with language, clothing, and sophisticated tools, took over Africa, then migrated beyond its borders. Certainly it's possible that this wave of human immigrants was spurred by mass deaths in the wake of the Toba eruption. But that's debatable.

What's certain is another explosion that nobody denies: the one in human symbolic communication. Our capacity for culture is what allowed us to survive in the perilous lands beyond the warm, fecund West African regions where *Australopithecus* first stood up. We never stayed in any one place for long. We moved into new places, founding new communities. And when we evolved complex symbolic intelligence, our growing facility with tools and language made these migrations easier. We could take advantage of many kinds of environments, teaching each other about their bounties and dangers in advance.

As *H. sapiens* poured off the continent of our birth, we discovered lands inhabited by our sibling hominins. We had to adapt to a world that already had humans in it. What came next will take us into one of the most controversial areas of population genetics and human evolutionary history.

7. MEETING THE NEANDERTHALS

NEANDERTHALS WERE HUMANS who went extinct between 20,000 and 30,000 years ago. Though there is some debate about who these people were, there is no question that there are none left. All that remains of the hundreds of Neanderthal groups that roved across Europe and Central Asia are a handful of ambiguous funeral sites, bones, tools, and pieces of art–along with some DNA that modern humans inherited from them. How can we avoid meeting the Neanderthals' fate? That depends on what you think wiped out these early humans in the millennia after they met *H. sapiens*.

By 40,000 years ago, humans had spread in waves across most of the world, from Africa to Europe, Asia, and even Australia. But these humans were not all perfectly alike. When some groups of *H. sapiens* poured out of Africa, they walked north, then west. In this thickly forested land, they came face-to-face with other humans, stockier and lighter skinned than themselves, who had been living for thousands of years in the cold wilds of Europe, Russia, and Central Asia. Today we call these humans Neanderthals, a name derived from the Neander Valley caves in Germany where the first Neanderthal skull was identified in the nineteenth century.

Neanderthals were not one unified group. They had spread far enough across Europe, Asia, and the Middle East that they formed regional groups, something like modern human tribes or races, who probably looked fairly different from each other. Neanderthals used tools and fire, just as *H. sapiens* did, and the different Neanderthal groups probably had a variety of languages and cultural traditions. But in many ways they were dramatically unlike *H. sapiens,* leading isolated lives in small bands of 10 to 15 people, with few resources. They had several tools, including spears for hunting and sharpened flints for scraping hides, cutting meat, and cracking bones. Unlike *H. sapiens,* who ate a wide range of vegetables and meat, Neanderthals were mostly meat-eaters who endured often horrifically difficult seasons with very little food. Still, there is evidence that they cared for each other through hardship: fossils retrieved from a cave in Iraq include the skeleton of a Neanderthal who had been terribly injured, with a smashed eye socket and severed arm, whose bones had nevertheless healed over time. Like humans today, these hominins nursed each other back to health after life-threatening injuries.

Roughly 10,000 years after their first meeting with *H. sapiens,* all the Neanderthal groups were extinct and *H. sapiens* was the dominant hominin on Earth. What happened during those millennia when *H. sapiens* lived alongside creatures who must have looked to them like humanoid aliens?

A few decades ago, most scientists would have answered that it was a nightmare. Stanford's Richard Klein, who spent years in France comparing the tools of Neanderthals and early *H. sapiens,* lowered his voice a register when I recently asked him to describe the meeting between these hominin groups. "You don't like to think about a holocaust, but it's quite possible," he said. He referred to the long-standing belief among many anthropologists that *H. sapiens* exterminated Neanderthals with superior weapons and intellect. For a long time, there seemed to be no other explanation for the rapid disappearance of Neanderthals after *H. sapiens* arrived in their territories.

Today, however, there is a growing body of evidence from the field of population genetics that tells a very different story about what happened when the two groups of early humans lived together, sharing the same caves and hearths. Anthropologists like Milford Wolpoff, of the

University of Michigan, and John Hawks have suggested that the two groups formed a new, hybrid human culture. Instead of exterminating Neanderthals, their theory goes, *H. sapiens* had children with them until Neanderthals' genetic uniqueness slowly dissolved into *H. sapiens* over the generations. This idea is supported by compelling evidence that modern humans carry Neanderthal genes in our DNA.

Regardless of whether *H. sapiens* murdered or married the Neanderthals they met in the frozen forests of Europe and Russia, the fact remains that our barrel-chested cousins no longer walk among us. They are a group of humans who went extinct. The story of how that happened is as much about survival as it is about destruction.

The Neanderthal Way of Life

We have only fragmentary evidence of what Neanderthal life was like before the arrival of *H. sapiens*. Though they would have looked different from *H. sapiens,* they were not another species. Some anthropologists call Neanderthals a "subspecies" to indicate their evolutionary divergence from us, but there is strong evidence that Neanderthals could and did interbreed with *H. sapiens*. Contrary to popular belief, Neanderthals probably weren't swarthy; it's likely that these early humans were pale-skinned, possibly with red hair. We know that they used their spears to hunt mammoths and other big game. Many Neanderthal skeletons are distorted by broken bones that healed, often crookedly; this suggests that they killed game in close combat with it, sustaining many injuries in the process. They struggled with dramatic climate changes too. The European and Asian climates swung between little ice ages and warmer periods during the height of Neanderthal life, and these temperature changes would have constantly pushed the Neanderthals out of familiar hunting grounds. Many of them took shelter from the weather in roomy caves overlooking forested valleys or coastal cliffs.

Though their range extended from Western Europe to Central Asia, the Neanderthal population was probably quite small—a generous estimate would put it at 100,000 individuals total at its apex, and many scientists believe it could have been under 10,000. By examining the growth of enamel on Neanderthal teeth, anthropologists have determined that many suffered periods of extreme hunger while they were young. This problem may have been exacerbated by their meat-heavy diets. When mammoth hunting didn't go well, or a particularly cold season left their favored game skinny or sick, the Neanderthals would have gone through months of malnutrition. Though Neanderthals buried their dead, made tools, and (at least in one case) built houses out of mammoth bones, we have no traditional evidence that they had language or culture as we know them. Usually such evidence comes in the form of art or symbolic items left behind. Neanderthals did make art and complex tools after meeting *H. sapiens,* but we have yet to find any art that is unambiguously Neanderthal in origin.

Still, there are intriguing hints. A 60,000-year-old Neanderthal grave recently discovered in Spain suggests that Neanderthals may have had symbolic communication before *H. sapiens* arrived. Researchers discovered the remains of three Neanderthals who appeared to have been gently laid in identical positions, their arms raised over their heads, then covered in rocks. The severed paws of a panther were found with the bodies, heightening the impression that the discovery represented a funeral ritual complete with "burial goods," or symbolic items placed in the graves. Erik Trinkhaus, an anthropologist at Washington University in St. Louis, says this site shows that Neanderthals might have had symbolic intelligence like modern humans.

Gravesites like these have led many scientists, including Trinkhaus, to believe that Neanderthals talked or even sang. But we haven't found enough archaeological evidence to sway the entire scientific community one way or the other.

By contrast, the *H. sapiens* groups who lived at the time of first contact with Neanderthals left behind ample evidence of symbolic thought. Bone needles attest to the fact that *H. sapiens* sewed clothing, and pierced shells suggest jewelry. There are even traces of red-ochre mixtures found in many *H. sapiens* campsites, which could have been used for anything from paint or dye to makeup. Added together, these bits of evidence suggest that *H. sapiens* groups weren't just using tools for survival; they were using them for adornment. And culture as we know it probably started with those simple adornments.

Looked at from the perspective of Neanderthals, then, there might have been a vast gulf between themselves and the newly arrived *H. sapiens*. The newcomers not only looked different–they were taller, slimmer, and had smaller skulls–but they probably chattered in an incomprehensibly complex language and wore bizarre garments. Would Neanderthals have tried to communicate with these people, or invited them to a dinner of mammoth meat?

For anthropologists like Klein, who spoke about a Neanderthal holocaust, the answer is an emphatic no. He's part of a school of anthropological thought that holds that *H. sapiens* would have met the Neanderthals with nothing but hate, disgust, and indifference to their plight. After those Neanderthals watched *H. sapiens* arrive, the next chapter in their lives would have been marked by bloodshed and starvation as *H. sapiens* murdered and outhunted them with their superior weaponry. Neanderthals were so poor, and had such a small population, that their extinction was inevitable.

This story might sound familiar to anyone versed in the colonial history of the Americas. It's as if *H. sapiens* is playing the role of Europeans arriving in their ships, and Neanderthals are playing that of the soon-to-be-exterminated natives. But Klein sees a sharp contrast between Neanderthals and the natives that Europeans met in America. When *H. sapiens* arrived, he asserted, "there was no cultural exchange" because the Neanderthals had no culture. Imagine what might have happened if the Spanish had arrived in the Americas, but the locals had no wealth, science, sprawling cities, nor vast farms. The Neanderthals had nothing to trade with *H. sapiens*, and so the newcomers saw them as animals. Neanderthals may have had fleeting sexual relationships with *H. sapiens* here and there, admitted Klein, but "modern human males will mate with anything." Tattersall agreed. "Maybe there was some Pleistocene hanky-panky," he joked. But it wasn't a sign of cultural bonding. For anthropologists like Klein and Tattersall, any noncombative relationships forged between the two human groups were more like fraternization than fraternity.

But there is a counter-narrative told by a new generation of anthropologists. Bolstered by genetic discoveries that have revealed traces of Neanderthal genes in the modern human genome, these scientists argue that there was a lot more than hanky-panky going on. Indeed, there is evidence that the arrival of *H. sapiens* may have dramatically transformed the impoverished Neanderthal culture. Some Neanderthal cave sites hold a mixture of traditional Neanderthal tools and *H. sapiens* tools. It's hard to say whether these remains demonstrate an evolving hybrid culture, or if *H. sapiens* simply took over Neanderthal caves and began leaving their garbage in the same pits that the Neanderthals once used. Still, many caves that housed Neanderthals shortly before the group went extinct are full of ornaments, tools, and even paints. Were they emulating their *H. sapiens* counterparts? Had they become part of an early human melting pot, engaging in the very cultural exchange that Klein and Tattersall have dismissed?

Extermination and Assimilation

The complicated debate over what happened to Neanderthals can be boiled down to two dominant theories: Either *H. sapiens* destroyed the other humans, or joined up with them.

The "African replacement" theory, sometimes called the recent African origins theory, holds that *H. sapiens* charged out of Africa and crushed *H. neanderthalensis* underfoot. This fits with Klein's account of a Neanderthal holocaust. Basically, *H. sapiens* groups replaced their distant cousins, probably by making war on them and taking over their territories. This theory is simple, and has the virtue of matching the archaeological evidence we find in caves where Neanderthal remains are below those of *H. sapiens*, as if modern humans pushed their Neanderthal counterparts out into the cold to die.

In the late 1980s, a University of Hawaii biochemist named Rebecca Cann and her colleagues found a way to support the African replacement theory with genetic evidence, too. Cann's team published the results of an exhaustive study of mitochondrial DNA, small bits of genetic material that pass unchanged from mothers to children. They discovered that all humans on Earth could trace their genetic ancestry back to a single *H. sapiens* woman from Africa, nicknamed Mitochondrial Eve. If all of us can trace our roots back to one African woman, then how could we be the products of crossbreeding? We must have rolled triumphantly over the Neanderthals, spreading Mitochondrial Eve's DNA everywhere we went. But mitochondrial DNA offers us only a small part of the genetic picture. When scientists sequenced the full genomes of Neanderthals, they discovered several DNA sequences shared by modern humans and their Neanderthal cousins.

Besides, how likely is it that a group of *H. sapiens* nomads would attack a community of Neanderthals? These were explorers, after all, probably carrying their lives on their backs. Neanderthals may not have had a lot of tools, but they did have deadly spears they used to bring down mammoths. They had fire. Even with *H. sapiens*' greater numbers, would these interlopers have had the resources to mount a civilization-erasing attack? Rather than starting a resource-intensive war against their neighbors, many *H. sapiens* could have opted to trade with the odd-looking locals, and eventually move in next to them. Over time, through trade (and, yes, the occasional battle) the two groups would have shared so much culturally and genetically that it would become impossible to tell them apart.

This is precisely the kind of thinking that animates what's called the multiregional theory of human development. Popularized by Wolpoff and his colleague John Hawks, this theory fits with the same archaeological evidence that supports the African replacement theory–it's just a very different interpretation.

Wolpoff's idea hinges on the notion that the ancestors of Neanderthals and *H. sapiens* didn't leave Africa as distinct groups, never to see each other again until the fateful meeting that Klein described with such horror. Instead, Wolpoff suggests, humans leaving Africa 1.8 million years ago forged a pathway that many other archaic humans walked–in both directions. Instead of embarking on several distinct migrations off the continent, humans expanded their territories little by little, essentially moving next door to their old communities rather than trekking thousands of kilometers to new homes. Indeed, the very notion of an "out of Africa" migration is based on an artificial political boundary between Africa and Asia, which would have been meaningless to our ancestors. They expanded to fill the tropical forests they loved, which happened to stretch across Africa and Asia during many periods in human evolution. Early humans would have been drifting back and forth between Africa, Asia, and Europe for hundreds of thousands of years. It was all just forest to Neanderthals and *H. sapiens*.

If scientists like Wolpoff are right—and Hawks has presented compelling genetic evidence to back them up—then *H. sapiens* probably didn't march out of Africa all at once and crush all the other humans. Instead, they evolved all over the world through an extended kinship network that may have included Neanderthals as well as other early humans like Denisovans and *H. erectus*.

It's important to understand that the multiregional theory does *not* suggest that two or three separate human lineages evolved in parallel, leading to present-day racial groups. That's a common misinterpretation. Multiregionalism describes a human migration scenario similar to those we're familiar with among humans today, where people cross back and forth between regions all the time. For multiregionalists, there were never two distinct waves of immigration, with one leading to Neanderthals, and the other packed with *H. sapiens* hundreds of thousands of years later. Instead, the migration (and evolution) of *H. sapiens* started 1.8 million years ago and never stopped.

Many anthropologists believe that the truth lies somewhere in between African replacement and multiregionalism. Perhaps there were a few distinct waves of migration, such anthropologists will concede, but *H. sapiens* didn't "replace" the Neanderthals. Instead, *H. sapiens* bands probably assimilated their unusual cousins through the early human version of intermarriage.

Perhaps, when Neanderthals stood in the smooth stone entries to their caves and watched *H. sapiens* first entering their wooded valleys, they saw opportunity rather than a confusing threat. In this version of events, our ancient human siblings may have had few resources and lived a hardscrabble life, but they were *H. sapiens*' mental equals. They exchanged ideas with the newcomers, developed ways of communicating, and raised families together. Their hybrid children deeply affected the future of our species, with a few of the most successful Neanderthal genes drifting outward into some of the *H. sapiens* population. Neanderthals went extinct, but their hybrid children survived by joining us.

Whether you believe that humans exterminated or assimilated Neanderthals depends a lot on what you believe about your own species. Klein doesn't think Neanderthals were inferior humans doomed to die—he simply believes that early *H. sapiens* would have been more likely to kill and rape their way across Europe in a Neanderthal holocaust, rather than making alliances with the locals. As his comment about the sexual predilections of modern men makes clear, Klein is basing his theory on what he's observed of *H. sapiens* in the contemporary world. Tattersall amplified Klein's comments by saying that he thinks humans 40,000 years ago probably treated Neanderthals the way we treat each other today. "Today, *Homo sapiens* is the biggest threat to its own survival. And [the Neanderthal extinction] fits that picture," he said. Ultimately, Tattersall believes that we wiped out the Neanderthals just the way we're wiping ourselves out today.

Hawks, on the other hand, described a more complicated relationship between *H. sapiens* and Neanderthals. He believes that Neanderthals had the capacity to develop culture, but simply didn't have the resources. "They made it in a world where very few of us would make it," he said, referring to the incredible cold and food scarcity in the regions Neanderthals called home. Anthropologists, according to Hawks, often ask the wrong questions of our extinct siblings: "Why didn't you invent a bow and arrow? Why didn't you build houses? Why didn't you do it like we would?" He thinks the answer isn't that the Neanderthals couldn't but that they didn't have the same ability to share ideas between groups the way *H. sapiens* did. Their bands were so spread out and remote that they didn't have a chance to share information and adapt their tools to life in new environments. "They were different, but that doesn't mean there was a gulf between us," Hawks concluded. "They did things working with constraints that people today have trouble understanding." Put another way, Neanderthals spent all day in often fatal battles to get enough food for their kids to eat. As a result, they didn't have the energy to invent bows and arrows in the evening. Despite these limitations, they formed their small communities, hunted collectively, cared for each other, and honored their dead.

When *H. sapiens* arrived, Neanderthals finally had access to the kind of symbolic communication and technological adaptations they'd never been able to develop before. Ample archaeological evidence shows that they quickly learned the skills *H. sapiens* had brought with them, and started using them to adapt to a world they shared with many other groups who exchanged ideas on a regular basis. Instead of being driven into extinction, they enjoyed the wealth of *H. sapiens*' culture and underwent a cultural explosion of their own. To put it another way, *H. sapiens* assimilated the Neanderthals. This process was no doubt partly coercive, the way assimilation so often is today.

More evidence for Hawks's claims comes from Neanderthal DNA. Samples of their genetic material can reveal just what happened after all that Pleistocene hanky-panky. A group of geneticists at the Max Planck Institute, led by Svante Pääbo, sequenced the genomes of a few Neanderthals who had died less than 38,000 years ago. After isolating a few genetic sequences that appear unique to Neanderthals, they found evidence that a subset of these sequences entered the *H. sapiens* genome after the first contact between the two peoples. Though this evidence does not prove definitively that genes flowed from Neanderthals into modern humans, it's a strong argument for an assimilationist scenario rather than extermination.

A big question for anthropologists has been whether *H. sapiens* comes from a "pure" lineage that springs from a single line of hominins like Mitochondrial Eve. The more the genetic evidence piles up, however, the more likely it seems that our lineage is a patchwork quilt of many peoples and cultures who intermingled as they spread across the globe. Present-day humans are the offspring of people who survived grueling immigrations, harsh climates, and Earth-shattering disasters.

Most anthropologists are comfortable admitting that we just don't know what happened when early humans left Africa, and are used to revising their theories when new evidence presents itself. Klein's influential textbook *The Human Career* is full of caveats about how many of these theories are under constant debate and revision. In 2011, for example, the anthropologist Simon Armitage published a paper suggesting that *H. sapiens* emerged from Africa as early as 200,000 years ago, settling in the Middle East. This flies in the face of previous theories, which hold that *H. sapiens* didn't leave Africa until about 70,000 years ago. The story of how our ancestors emerged from their birthplaces in Africa turns out to be as complicated as a soap opera—and it likely includes just as much sex and death, too.

Who Survived to Tell the Tale?

Whether humans destroyed Neanderthals or merged with them, we're left with a basic fact of anthropological history, which is that modern humans survived and Neanderthals did not. It's possible that members of *H. sapiens* were better survivors than their hominin siblings because Neanderthals didn't exchange symbolic information; they were too sparse, spread out, and impoverished to achieve a cultural critical mass the way their African counterparts did. But it seems that Neanderthals were still swept up into *H. sapiens'* way of life in the end. Our Neanderthal siblings survive in modern human DNA because they formed intimate bonds with their new human neighbors.

Svante Pääbo, who led the Neanderthal DNA sequencing project, recently announced a new discovery that also sheds light on why *H. sapiens* might have been a better survivor than *H. neanderthalensis*. After analyzing a newly sequenced genome from a Denisovan, a hominin more closely related to Neanderthals than *H. sapiens* are, Pääbo's team concluded that there were a few distinct regions of DNA that *H. sapiens* did not share with either Neanderthals or Denisovans. Several of those regions contain genes connected to the neurological connections that humans can form in their brains. In other words, it's possible that *H. sapiens'* greater capacity for symbolic thought is connected to unique strands of DNA that the Neanderthals didn't have.

"It makes a lot of sense to speculate that what had happened is about connectivity in the brain, because . . . Neanderthals had just as large brains as modern humans had," Pääbo said at a press conference in 2012 after announcing his discovery. "Relative to body size, they had even a bit larger brains [than *H. sapiens*]. Yet there is something special that happens with modern humans. It's sort of this extremely rapid technological cultural development and large societal systems, and so on." In other words, *H. sapiens*' brains were wired slightly differently than their fellow hominins. And once Neanderthals merged with *H. sapiens*' communities, bearing children with the new arrivals, their mixed offspring may have had brains that were wired differently, too. Looked at in this light, it's as if *H. sapiens* assimilated Neanderthals both biologically and culturally into an idea-sharing tradition that facilitated rapid adaptation even to extremely harsh conditions.

Early humans evolved brains that helped us spread ideas to our compatriots even as we scattered to live among new families and communities. It's possible that this connectedness—both neurological and social—is what allowed groups of *H. sapiens* to assimilate their siblings, the Neanderthals. Still, our storytelling abilities are also what allow us to remember these distant, strange ancestors today.

Humans' greatest strength 30,000 years ago may have been an uncanny ability to assimilate other cultures. But in more recent human history, this kind of connectedness almost did us in. Once human culture scaled up to incorporate unprecedentedly enormous populations, our appetite for assimilation spread plagues throughout the modern world, almost destroying humanity many times over. And it spawned deadly famines, too. As we'll see in the next two chapters, humanity's old community-building habits can become pathological on a mass scale. Thousands of years after the merging of Neanderthals and *H. sapiens*, the practices that helped us survive in pre–ice age Europe became, in some contexts, liabilities. They wiped out whole civilizations and made it necessary for us to change the structures of human community forever.

8. GREAT PLAGUES

[Death] hath a thousand slayn this pestilence.
And, maister, er ye come in his presence,
Me thinketh that it were necessarie
For to be war of switch an adversarie.
Beth redy for to meete hym everemoore.

—Geoffrey Chaucer, "The Pardoner's Tale," *The Canterbury Tales*, 1380s

MOST PEOPLE KNOW British poet Geoffrey Chaucer because he wrote one of the earliest works of literature in English, *The Canterbury Tales*. What you may not know is that Chaucer came of age in a postapocalyptic world. Born in the 1340s, Chaucer would have been a little boy when the Black Death first struck England, in 1348; in the next few years it wiped out over 60 percent of the population of the British Isles. The son of a wealthy wine merchant, Chaucer grew up in London, already a bustling city where traders arriving in ships from Europe would have brought news of the "pestilence" ripping through the continent. The late 1340s marked the first great pandemic of what would later be called the bubonic plague, and the death tolls were so high that most bodies were thrown into mass graves because churchyards were overflowing. Even if there had been room for the corpses, it's likely there were not enough clergy left to coordinate burials. Chaucer came of age in the wake of a pandemic so deadly that half the population of London perished.

We hear of the Black Death only rarely in Chaucer's considerable body of work, most memorably in the lines I've quoted above from *The Canterbury Tales*. The corrupt Pardoner is telling his fellow travelers a tale of three angry drunks who decide to kill Death, to avenge their friend's murder. Violently intoxicated, they demand that a little boy carrying corpses to the graveyard tell them where to find "Deeth." The boy warns them that Death "hath a thousand slayne this pestilence," or slain a thousand people during the last bout of plague. The boy adds that they should be ready to meet Death "everemoore," anytime. This casual reference to "pestilence," written over 40 years after the plague first shattered England, indicates how ordinary the specter of mass death had become for people of Chaucer's generation. The disease had returned again and again to claim thousands of lives during the late fourteenth century, though not with the ferocity that it had in Chaucer's boyhood. The pestilence may not have touched the poet's writings much, but its social reverberations marked his life and those of all his countrymen.

Plague was a symptom of the problems humans had adapting to our own growing societies. By the time the Middle Ages rolled around, we were old pros at the symbolic-culture game that helped us outlast the Neanderthals, but we still had little experience with using that symbolic culture to unite large societies made up of many disparate groups. Humans first began experimenting with such societies during classical antiquity, in sprawling ancient empires like those of the Assyrians, the Romans, the Han, and the Inca. But these civilizations were exceptions rather than the norm for most people. During Chaucer's lifetime in the Middle Ages, however, humanity began laying the foundations for what would over the next five centuries grow into modern, global society. And this transformation meant that for the first time, the greatest threat to humanity came not from nature, but from ourselves.

A Revolutionary Pestilence

In a country whose population was only 40 percent of what it had been in the years before his birth, Chaucer grew up with opportunities he might never have had otherwise. A man of lively intelligence, he got an uncommonly good education working as an esquire at the court of Edward III, a typical role for the child of a wealthy merchant. He managed to get good legal training by studying with attorneys who worked in the court's "Inner Temple," essentially a medieval law school. And then he found paying work as a representative for various members of the royal family, conducting business for them abroad (where he learned French and Italian) and eventually in London. Because Chaucer did so much business for the crown, he left a surprisingly detailed paper trail, including travel authorizations, expense accounts, promises of payment, and legal documents. Scholars have pieced together his life from these scraps of paper.

We know from these records that for twelve years, the former esquire lived with his wife and children in Aldgate, one of the wealthiest neighborhoods of London. Chaucer's home was a fine set of rooms right above a gate in the ancient defensive walls that surrounded the city. In typical feudal fashion, the mayor had granted the dwelling to Chaucer rent-free at roughly the same time that Edward III put the future poet in charge of managing export taxes on wool in London's Custom House. Apparently, Chaucer was quite good at his job. He did valuable accounting for the kingdom during the day, and probably made his first efforts at writing poetry during the evenings. Though Chaucer managed vast sums of money for the crown, he and his family were what would one day be called middle class. Connections to the royal family gave them just enough stature to merit a good living (his salary included a daily gallon pitcher of wine), and a nice home. One side effect of the Black Death was a dramatic reshuffling in the upper echelons of society, whose members had been thinned by the pestilence. Chaucer flitted from one good job to the next, always working closely with the crown, because there was a shortage of educated men who did not owe blood allegiance to one aristocratic family or another.

The same grisly population crash that sent Chaucer and his family up the social ladder was affecting the peasant class, too. And in 1381, Chaucer's life was threatened by the results. It was the year of the Peasants' Revolt, a series of violent riots fomented by peasants demanding better wages and treatment. They happened literally beneath the Chaucer family's windows in Aldgate, and many of the rioters were armed with weapons and torches; the angry protesters left a lot of Chaucer's rich neighbors dead.

But why did a disease epidemic lead to riots for better pay? Call it cul-

An image from the British Library (circa 1385–1400), depicting the Mayor of London executing Wat Tyler, leader of the Peasants' Revolt, while King Richard II looks on.

tural adaptation in action. Jo Hays, a historian at Loyola University whose work focuses on pandemics in the ancient world and the Middle Ages, explained that the Black Death had upset a stalemate in class relations that had lasted centuries. Peasants, long tied to the land by the feudal system, had been trapped in serfdom because there were no other options. As the peasant population ballooned, their lords could afford to grant them less and less—it was a landlord's market, as it were. The poor starved, but had no options. When the Black Death hit, most of the people who died were impoverished folk whose health was already compromised by lack of food. The survivors, like Chaucer, found themselves in a world where jobs were suddenly plentiful. Couple this situation with the rise of money wages (like those Chaucer's family got as merchants) instead of land grants, and for the first time in centuries, peasants could exercise a degree of choice over their work.

The magnitude of the plague also called into question every form of authority in the medieval world. Though the Church called the pestilence a punishment from God, it was hard to avoid noticing that so-called godly men and women were meeting the same fate as the godless–nor did their prayers prevent the plague. Indeed, Chaucer was highly critical of Church officials in *The Canterbury Tales,* a sentiment he voiced because it reflected popular, though controversial, ideas in his day. Likewise, the common people began questioning government authorities in the wake of the Black Death, especially once serfs had a better bargaining position.

Still, those authorities did their best to cling to the old rules. In the wake of the Black Death, the British government reacted to the labor shortage by trying to limit wages by law. In 1351, just four years after the first plague year, the English Statute of Laborers stipulated that all wages should be held at the pre-pestilence levels of 1340. The situation quickly became untenable. Angry peasants demanding better wages stormed Chaucer's neighborhood, burning homes and dragging rich people from their rooms to be killed. It's not clear whether Chaucer was actually at home when the riots happened, but some of his friends and business associates were murdered during the mob violence. Soon after, the government repealed the statute, along with similar laws, allowing peasants to earn higher wages and achieve some autonomy from their landlords. Within days of the Peasants' Revolt, Chaucer gave up his cozy rooms in Aldgate and moved permanently to Kent, near Canterbury, beginning a new career as an estate manager and city planner. The city where he'd grown up had transformed so radically that he no longer felt welcome there.

Similar rebellions racked France and Italy, as the surviving laborers realized how valuable they were in a depopulated Europe. Though bloody at first, this kind of rebellion led to better treatment of workers and eventually to the rise of the middle classes. The benefits of these social reforms extended even to women, a group who had almost never been part of the traditional labor force. After the Black Death, there was a rise in the number of women running their own taverns. (We see the evidence in records of who was purchasing grain for brewing beer, where there is a notable uptick in female buyers.)

The first wave of plague also led directly to some of the first city-planning efforts aimed at improving the lives of the general populace. In the wake of the pandemic, many cities established boards of public health, which by the fifteenth century were responsible for sanitation and waste disposal in cities like Florence and Milan. These boards also engaged in what today we'd call "health surveillance," compiling weekly lists of people who died from epidemic diseases so that officials could spot a pandemic before it became widespread. Just as we do today, city officials in early Renaissance Florence would establish quarantines for people afflicted by disease, to prevent a major outbreak.

"The Poor That Cannot Be Taken Notice Of"

The scale of the fatalities from the Black Death during the 1300s was caused by the structure of the societies where it spread. Close living conditions made a deadly disease into an apocalyptic pandemic. SUNY Albany anthropologist Sharon DeWitte worked with a group of biologists who sequenced bacterial DNA from medieval victims of the epidemic. They discovered ample evidence that bubonic plague was caused by the bacterium *Yersinia pestis*, which is almost always spread by contact, not air. Urban societies, with their closely packed populations, were therefore ideal breeding grounds for *Y. pestis*. The other major cause of the near-extinction events during Chaucer's childhood was lack of food. In 1348, the Black Death came soon after a terrible famine had weakened the immune systems of people, mostly the poor, who had the least to eat. An epidemic became a "Black Death" in part because of how the ruling class allocated economic resources.

The plagues of the late fourteenth century called attention to one of the greatest threats to human adaptability in an urbanizing world. Put in the simplest possible terms, that threat is a stark class division between rich and poor. When many people live in close proximity, but a large portion of them are trapped in deprived, unhealthy conditions, the entire society is put at risk of extinction. Pandemics spread rapidly among the vulnerable, bringing death to everyone. But this isn't just a matter of epidemiology. Feudalism was an economic system which kept a major part of the population locked into poverty, and it was so rigid that any perturbation of the social order left it open to disruption. The pestilence that Chaucer described coming "everemoore" attacked a society whose rules made it both biologically and culturally vulnerable.

And yet the humans who survived one of the greatest disease apocalypses in our history did not respond with despair and a descent into savagery. There was no zombie freakout scenario, as we like to imagine today. Instead, the Peasants' Revolt led to social reforms that improved the lot of the poor in the decades that followed. Our facility for cultural adaptation can bloom even in the wake of seeming apocalypse. Though it would be centuries before the renewed interest in science that arose with the Age of Enlightenment, let alone the germ theory of disease, the humans who survived the plague managed to lay the foundations for political structures in which every class could advocate for its own best interests. At the same time, newly created health boards stood a chance of protecting vulnerable populations, too.

As European cities grew and feudalism crumbled, the rise of the market economy forged new connections between urban societies through international trade. Humans once again raced to adapt to the dangers created by a global civilization with a massive, vulnerable population.

Though epidemics seemed to hit roughly once a generation in England, the plague summer of 1665 killed so many people that it was called simply the Great Plague. It was also a period of rapid cultural change, when class divisions again took on deadly proportions. In the diaries of a young, well-connected naval official named Samuel Pepys (pronounced *peeps*), we have a record of daily life during this time.

In late August of 1665, Pepys described the streets of London in his diary:

> But now, how few people I see, and those walking like people that have taken leave of the world.... Thus this month ends, with great sadness upon the public through the greatness of the plague, everywhere through the Kingdom almost. Every day sadder and

> sadder news of its increase. In the City died this week 7496; and all of them, 6102 of the plague. But it is feared that the true number of the dead this week is near 10000—partly from the poor that cannot be taken notice of through the greatness of the number, and partly from the Quakers and others that will not have any bell ring for them. As to myself, I am very well; only, in fear of the plague.

Pepys's horror grows as the death tolls rise, even as he must continue going about his business—which, by the way, was booming in the plague year. As his neighbors fall prey to the disease, he sees "Searchers," groups mostly of women, who inspect houses looking for evidence of the Black Death. Where they find it, the Searchers impose quarantine on the people living there, and mark the doors with red crosses. In the passage I've excerpted from above, Pepys describes empty streets where a few brave people outside like himself walk around in a daze. He's wandering through an apocalyptic urban landscape, recognizable to anyone who has watched movies like *28 Days Later* or seen the TV series *The Walking Dead*.

Without access to medicine, crowded together in densely packed slums, the London poor succumbed to plague swiftly. New York University's literary historian Ernest Gilman has pored over writings from this era, where representatives of the Church insisted that the Black Death was a punishment from God. But, he noted, by the seventeenth century these men were in dialogue with a group sneeringly referred to as "mere naturians," or proto-scientific thinkers who believed the plague had a purely earthly origin. Though most medicine at the time would be called quackery today, the official government position was nevertheless that the disease was contagious. It was said to spread through "the miasma," the air. These ideas led to the practice of state-enforced quarantining, but also to people wearing face masks and even washing coins in vinegar when they changed hands. Medicine and science were ideas that had achieved some social currency during Pepys's time; a lot had changed since Chaucer dared to make fun of the Catholic Church in *The Canterbury Tales*.

What had also changed was the marketplace. A new class of merchants and tradespeople had come to occupy a central place in England and Europe's economic systems, and they established trade routes throughout the world. Cities became central to this new economy, and impoverished groups flocked to the slums of big cities like London, hoping to find their fortunes in the world of trade rather than farming. As a result, Pepys could observe the stark class division between those touched by plague and those decimated by it. Involved as he was in naval matters and trade, Pepys could also profit from the very market systems that most helped set up conditions for the Great Plague. Tragically, the first stirrings of global capitalism and disease seemed to go hand in hand.

In his diary, Pepys also noted something that's crucial for understanding the spread of epidemics during the seventeenth century and beyond: Mortality rates among the poor were skyrocketing, and yet at the same time were not being recorded. He wrote that many believed the death toll was likely 2,500 more people than officially reported, "partly from the poor that cannot be taken notice of through the greatness of the number."

Nothing would make that more obvious than the devastating epidemics that were sweeping the Americas while Pepys was getting rich back in London.

The Plagues of Colonialism

In Pepys's time, Europeans had been carving out colonies in the Americas for over a century and a half. A lucrative trade in goods and people turned the Atlantic into a maze of shipping lanes, packed with cargo vessels bearing everything from gold and slaves to animals and produce. They also bore disease.

One of the enduring questions in American history is why ragtag groups of European and English colonists were able, in just a couple of centuries, to claim the riches of two continents packed with enormous cities in the Aztec and Inca empires, along with highly trained armies, vast farms, and millions of people. In the seventeenth century, the dominant theory would have been that God was dishing out justice to the heathen natives. Up until the mid-twentieth century, historians and anthropologists offered rationales that weren't much better. They believed the natives were too innocent, savage, stupid, or inferior to mount a decent defense against the European invaders. In the late 1990s, Jared Diamond argued in *Guns, Germs, and Steel* that the Inca weren't culturally inferior, but instead victims of historical and environmental circumstances. Diamond popularized the idea that the Inca fell to the Spanish because the Americans' "stone age" technology and lack of writing left them unprepared to deal with the Europeans' guns, cavalry, and greater stores of knowledge. These issues, as much as the plagues Europeans brought, were what left the Inca empire vulnerable to conquest.

Over the past decade, however, new information has emerged about the civilizations of the Americas. As Charles Mann explains in his book *1491*, an exploration of new scholarship on pre-Columbian life, the Inca were technologically advanced enough to have defeated the Spanish. They had a highly developed system of writing called quipu, created by making different kinds of knots out of string, which is only today finally being deciphered (sadly, the Spanish burned most of the quipu libraries). True, the Inca did not have horses or metal weapons, but they had textile technologies that allowed them to weave massive boats from rushes, hurl flaming rocks over great distances using slingshots, and of course they had the advantage of a hilly terrain that was nearly impossible for horses to climb. What felled the Inca was quite simply a plague on the scale of what Chaucer witnessed as a boy, coupled with a raging civil war caused by a power vacuum left when several Inca leaders succumbed to smallpox.

By the time the conquistadors arrived in full force, South America was already riddled with plagues that spread easily on the vast trade routes of the Inca and Aztec empires. Imagine if a group of warriors, armed to the teeth, had descended on London in the wake of the Peasants' Revolt. The city was depleted, and its inhabitants were squabbling violently over what to do next. It would have been easy for even a small band of foreign soldiers to step in and lay waste to the city. As Mann would have it, Europeans did not conquer the Americas with technology and writing. Instead, they inadvertently imported smallpox, influenza, and bubonic plague to the Americas, and those diseases destroyed native communities long before conquistadors like Francisco Pizarro could come in to claim victory over them.

Many of these new theories were first popularized in the historian Alfred Crosby's influential 1972 book *The Columbian Exchange: Biological and Cultural Consequences of 1492*. In it, Crosby argues that European and American meetings constituted a vast environmental experiment, in which plants, animals, and microbes that had been separated for sometimes millions of years were suddenly thrown together. Peas, squash, potatoes, tomatoes, maize, and other native American crops were brought back to Europe; horses, pigs, and cows were brought to the Americas. Syphilis returned from the New World in the bodies of explorers, and Europe's plagues arrived in the New World the same way.

Just as people in Pepys's London had no scientific way to respond to plague, neither did citizens of the great cities like Tenochtitlán and Cuzco, in regions that later became Mexico and Peru. Based mostly on written accounts of the period from explorers, as well as death records in missions, many historians and anthropologists believe that as much as 90 percent of the American population was eventually felled by epidemics. Arizona State University forensic archaeologist Jane Buikstra, who has studied the remains of people who lived before colonial contact in Mexico and Peru, believes that the Columbian plagues hit populations that were already vulnerable. In the bones of people born before Europeans arrived, she said, "you see evidence of warfare and malnutrition. Some groups were highly stressed, living in constrained, unhealthy conditions with a lot of garbage around them." Stressed groups would be more vulnerable to introduced disease—much the way the urban slum dwellers in seventeenth-century London were, or the starved peasants in the Middle Ages.

Unlike England, however, the Americas were in the process of being colonized by foreign powers. And that, according to the historian Paul Kelton, of the University of Kansas, may have meant the difference between the typical European epidemic death tolls (up to 60 percent in the wake of the Black Death) and typical American ones (up to 90 percent). Kelton has studied historical records of North American native cultures like the Cherokee, and believes that the social and economic upheavals caused by colonialism exacerbated the virulence of American plagues. Without the colonists' trade routes linking previously isolated groups and regions together, Kelton argues, epidemics wouldn't have spread as rapidly. The main vector of disease on these trade routes would have been slaves taken from the American population. Even though most people don't know about it today, the traffic in native American slaves was a thriving business in seventeenth-century America. Slavers played into already existing tensions between rival groups, encouraging the victors in battle to trade their captives with Europeans for goods ranging from guns and powder to horses and wool clothes. Slave raids, in turn, intensified warfare, and disrupted centuries-old patterns of hunting and farming. Groups decimated by slavery, their strongest warriors shipped overseas to sugar plantations in the West Indies, hid from their enemies in fortified villages and were unable to secure food supplies they needed. When smallpox hit these villages, the death tolls were stupendous.

Worse than the initial wave of epidemics was the fact that American groups had no time to recover from their losses. In England, after the Black Death hit, common people were able to continue in the same jobs and homes they'd had before the pestilence. In fact, as the Peasants' Revolt makes clear, many were able to demand a better way of life because their labor had become more valuable. But in the Americas, new colonial governments and militias used their power to force plague-weakened groups off their lands, or tempted them into trading their livelihoods away for guns and horses. Biological catastrophe was followed by political catastrophe, which led to the kinds of displacement and poverty that can correlate with high death tolls in epidemics. In many regions, missionaries would push native groups to live in missions where men and women were separated, thus ensuring that the population couldn't rebuild itself by forming new families and having children.

David S. Jones, a Harvard historian and medical doctor, sums up the issues succinctly in his influential paper "Virgin Soils Revisited":

> Any factor that causes mental or physical stress—displacement, warfare, drought, destruction of crops, soil depletion, overwork, slavery, malnutrition, social and economic chaos—can increase susceptibility to disease. These same social and environmental fac-

tors also decrease fertility, preventing a population from replacing its losses.

American epidemics were likely triggered by the same factors as the European ones. The main difference was that the rapid advance of colonial intrusions in the Americas prevented populations from recovering before they were hit with another wave of disease.

Survival and Survivance

What the plagues of the past 700 years reveal is that human mass societies can magnify the effects of threats that come from the environment, like disease. As our cultures grow, so, too, do our vulnerabilities to extinction. There are many failure modes when we try to adapt to our new circumstances as creatures who can no longer wander off to found a new community the way our ancestors did. Rigid class divisions and warfare are two such failure modes, and they are often accompanied by pandemics. As the University of Colorado history professor Susan Kent explains in her recent book about the 1918–19 flu epidemic—the worst in human history since Chaucer's time—this pandemic virus quickly became more virulent because it spread with the movements of soldiers during World War I. As waves of soldiers succumbed to the flu, new ones came to replace the dying. The virus always had fresh new hosts, who were generally being shipped across the globe to new locations where the flu would take hold. Like the Black Death, the 1918–19 pandemic led to reforms in health care and indirectly sparked several colonial rebellions reminiscent of the Peasants' Revolt.

We can excavate a grim survivor's lesson from the piles of bones these pandemics left behind. We are currently struggling to adapt to life in a global society, where the dangers of culture-saturated, densely populated cities have replaced the dangers of the wilderness. And we are adapting. With each plague, there arise social movements that inch us closer to economic equality and clarify what's required to take a scientific approach to public health.

The lingering social effects of the American plagues are nevertheless a reminder that there's a lot of work that remains to be done. Those waves of colonial-era pandemics helped usher in an era of economic inequality between colonizers and the colonized, undermining civilizations that had thrived for thousands of years. Anishinaabe author and University of New Mexico American Studies professor Gerald Vizenor argues that native cultures and peoples have survived throughout the Americas, though often in dramatically altered form. They've done it by maintaining communities, passing on stories to younger generations, and fighting for political sovereignty when they could.

Vizenor coined the term "survivance" to describe the practices of natives today who are connected to their cultural traditions, but also living them dynamically, reshaping them to suit life in a world forever changed by colonial contact. The difference between survival and survivance is the difference between maintaining existence at a subsistence level and leading a life that is freely chosen. As we contemplate the ways humanity will endure, it's worth keeping the idea of survivance in mind. One of the best things about *H. sapiens* is that we are more than the sum of our biological parts. We are minds, cultures, and civilizations. I don't mean to say that people can live on ideals alone: That's obviously stupid. But when we aspire to survive disaster, we are perhaps without realizing it aspiring to survive as independent beings. We aren't aiming for a form of survival that looks like slavery or worse.

The European and American plagues changed the world, both environmentally and economically. They also revealed a basic truth: Survival is cultural as well as biological. To live, we need food and shelter. To live autonomously, we must remember who we are and where we came from. As we'll see in the next chapter, this is especially the case when it comes to another one of the greatest threats to human survival: famine. Often, the regions most deeply stricken by hunger are also places where people have been deprived of social and economic power.

9. THE HUNGRY GENERATIONS

IT'S BEEN CALLED Black '47, the Great Irish Famine, and the Potato Famine. From 1845 to 1850, Ireland was hit with one of the most brutal famines of the nineteenth century after several annual harvests of potatoes were ruined by blight. Two million people died, and the harshness of the experience sent at least a million more to seek new homes in the United States and England. Though the Irish population was 8 million in 1841, today it hovers at roughly half of that. The country still, over 160 years later, has not recovered from the aftereffects of a disaster that changed not just Ireland but the entire way we conceive of famine.

Famines have been recorded in historical documents and religious books for almost as long as humans have been writing, and yet they are still among the most poorly understood causes of mass death among humans. The University College Dublin economist Cormac Ó Gráda has spent most of his career studying famine, and admitted to me that it's very difficult to say how many people die of starvation when "most people in famine die of diseases." In fact, he added, malnutrition is a bigger killer than generally believed because it leads to the scenarios we explored in the previous chapter, where people are more vulnerable to epidemic disease. It's only in the last few centuries that we find reliable, complete records of famine that scholars like Ó Gráda can use to piece together the events that lead to masses of people dying from want of the most basic resource: food.

The Potato That Starves Us

Before the events of Black '47, the dominant theory of famine came from the eighteenth-century demographer Thomas Malthus, who believed that epidemics and famines were natural "checks" on human populations to keep them in balance with resources available. From the Malthusian perspective, famines should come on a regular basis, especially when a population is outstripping its ability to subsist in a particular area. When the Irish Potato Famine first began to unfold, however, journalists covering the events realized that the Malthusian explanation wasn't adequate. This famine had its roots in politics. In the decades leading up to Black '47, industrialization had completely reshaped the Irish landscape. Lands that were once dotted with small farms devoted to a variety of crops were taken over by landlords who converted these farms to pastureland—there, they raised animals for export. Seeking a high-yield crop that could feed their families, peasants on the land that remained available for farming switched from grain crops to potatoes. Many of these peasants were working entirely for "conacres," or the right to grow their own subsistence food on a landlord's property. They were living hand-to-mouth, on potatoes, and had no cash to use if their crop failed. When the blight struck, the Irish poor lost both food and money at once. Even at the time it seemed likely that the famine, rather than being a natural "check" on the island's growing population, was the result of political and economic disaster (itself partly the result of Ireland's colonial relationship with Britain).

In the century and a half since people began to perceive the political underpinnings of famines, it has become commonplace to view them as caused primarily by economic problems. The Nobel Prize–winning economist Amartya Sen first advanced this theory in the 1980s, in his highly influential book *Poverty and Famines: An Essay on Entitlement and Deprivation*. There, Sen lays out the details of his famous theory. He explains that "entitlements" are avenues by which people acquire food, and famine is always the result of how those entitlements work in a particular society. Direct entitlements refer to subsistence methods of getting food, like growing potatoes. Indirect entitlements are avenues that allow people to get food from others, usually by earning money and buying it. Transfer entitlements are ways that people get food when they have neither direct nor indirect means—generally, from famine-relief groups. Sen's theory has allowed economists to diagnose the causes of famines by looking at the causes of "entitlement failures." In the case of Black '47, most Irish people suffered all three forms of entitlement failure.

But there is a missing piece in Sen's theory of entitlements, and it's one that is only going to become more important as we move into the next century. That piece is the environment. Evan Fraser, a geographer at the University of Guelph, in Ontario, researches food security and land use. He argues that the Irish Potato Famine reveals how poor environmental management can lead to mass death. He believes we should supplement Sen's theory of entitlements with an understanding of how "ecological systems are vulnerable to disruption." In other words, famines are undeniably the result of how we use (or abuse) our environments to extract food from them.

Many famines begin when economic or political circumstances encourage people to convert environments into what Fraser calls "specialized landscapes," good for nothing but growing a limited set of crops. In Ireland, for example, landlords pushed farmers to transform diverse regions into landscapes that could yield only one crop: the potato. Often, this kind of farming appears to work brilliantly in the short term. Black '47 was preceded by years of escalating productivity as Irish farmers converted grain crops to potatoes. Paradoxically, the wealth of the ecosystem meant that it was also precarious. Any change to these "specialized landscapes," whether a blight or a slight change in temperature or rainfall, can wipe out not just one farm but *every single farm*. What's bad for one potato farm is bad for all of them. As Fraser put it, "You get one bad year, and you're stuffed." In the case of the Irish famine, there were at least two bad years of blight before true famine set in, during 1847.

The ecosystem vulnerabilities leading to Black '47 could very well become common over the next century. Seemingly minor problems like temperature and rainfall changes could spell death for regions that depend on a single crop that is sensitive to such changes. The most immediate areas of concern lie in the breadbaskets of the American Midwest, a vast region of prairies that stretches from Saskatchewan to Texas. "Most societies are interested in grain, and if you think in terms of wheat, then you need a hundred and ten days of growing conditions," Fraser said. "You need weather that's not too hot and not too cold, and abundant but not excessive rainfall. If you get long periods of good weather, you don't realize there could be a problem. And then you get one bad year, and it all unravels very quickly."

During the dust-bowl famines of the 1930s, farmers saw a collapse of the Midwest ecosystem. And we're going to see a return to the dust bowl again. "We know from climate records that the Midwest experiences two-hundred- to three-hundred-year droughts. There are periods where we see centuries of below-average precipitation. The twentieth century was above average. We've had a long rhythm of good weather. But the next hundred years will be much drier. We've already seen droughts hitting in Texas. It's going to be hard to maintain productivity then." I spoke to Fraser in early 2012, before the worst drought in over 50 years hit the Midwest that summer, destroying crops and livelihoods. According to the weekly U.S. Drought Monitor that year, "About 62.3 percent of the contiguous U.S. (about 52.6 percent of the U.S. including Alaska, Hawaii, and Puerto Rico) was classified as experiencing moderate to exceptional drought at the end of August." Fraser's predictions for drought are already coming to pass, and he and his colleagues believe more is yet to come.

Does this mean we are witnessing the leading edge of a global famine? Yes and no. The issue here isn't that people are inevitably victims of their environment, nor that mild changes in the weather always lead to famine. Trouble comes when you see a growing group of people who are extremely poor, combined with a vulnerable ecosystem that's not diverse and therefore can't withstand any kind of climate change or pests. A failed crop is a tragedy, but it doesn't become a famine unless people don't have the money to buy food elsewhere.

In Black '47, the main way that people survived the famine was a method that spoke to both problems. They emigrated, moving themselves and their families away from a failing economy and a failing ecosystem. But during many famines, people don't have that option.

The War That Divides Us

There were many famines during the mid-twentieth century, and most of them were related to war. A lot of these deaths were probably from diseases exacerbated by conflict and malnutrition. War rationing and deprivation leave people weakened, vulnerable to dysentery and epidemic disease. But in the early 2000s, Newcastle University historian and demographer Violetta Hionidou found evidence of a terrifying period during the Axis occupation of Greece from 1941 to 1944, when 5 percent of the population died directly from want of food.

The Greek Army staged a highly successful resistance to Italian invasion in 1940, and the Greek premier refused to buckle under to Mussolini even after a protracted battle. Indeed, had it not been for the aid that Bulgaria and Germany gave to the Italians, it's likely that Greece would have held fast. Instead, after intense German bombing, Greece fell and was occupied by troops from Italy, Bulgaria, and Germany—all of whom were taking orders from Nazi commanders in Germany, and securing the public's docility through a Greek puppet government in Athens. In the wake of the occupation, England withdrew its support (the British had been aiding the Greek military) and set up a blockade to prevent supplies from reaching the country. Carved up into territories occupied by three different hostile nations, and cut off from its former allies, Greece was fragmented and intensely vulnerable.

That fragmentation—political, economic, and, in the case of the famine-stricken islands Syros, Mykonos, and Chios, geographical—is what led to the horrors that came next. A first wave of famine struck Athens, claiming as many as 300,000 lives after the German occupying forces requisitioned food and demanded that the Greek government pay the costs of the occupation. In one fell swoop, the people of Athens had lost what Sen would call direct and indirect entitlements, and the blockade prevented transfer entitlements from easing their suffering. Often, that is where the story of the Great Famine of Greece is said to end, with a few nods made to the fact that there were also poor harvests. But according to Hionidou, there was much more to the story than that.

First of all, the production levels from Greek farming did not actually dip below the norm. She found that the numbers historians have used to make this claim are entirely based on products that the occupying government collected tax on. But there was widespread resistance to paying tax, as well as the simple fact that people couldn't afford it. Therefore a lot of foodstuffs that Greece produced during the famine went untaxed and unrecorded. Still, Greek citizens relied for much of their food on imports and trade; remote island areas were especially dependent on imports. The occupying forces restricted people's movements to small areas, which meant that nearly the entire Greek population had to sneak around and participate in a black market for food. Of course, as Hionidou pointed out, "Those who couldn't afford the black market died."

And then there were those who had no access to the black market at all. On some islands, there was no way, physically, to sneak past the blockades and get to people selling food. On Syros, Mykonos and Chios, for example, people had to depend entirely on the food they produced to eat. And there simply wasn't enough of it. The mortality patterns Hionidou found during her research flew in the face of the traditional idea that death from starvation is rare. She pored over records from the time, amazed to discover that accounts from both the Axis side and the Greek side matched up. "Greek doctors were reporting the cause of death as starvation, and some could argue that they had good reason to report starvation to blame the occupying forces," she said. "But the occupying forces produce documents talking about starvation, too. They don't try to cover it up by saying it's disease. They're not denying it at all."

The only way that people survived these outbreaks of famine was to hold out until 1942, when the blockade was loosened up and food aid reached the Greek people. Some managed to escape the country into Hungary, while others got rich on the black market. But the artificial barriers that the occupation erected between people did more to starve them than any failed crop ever could. For Hionidou the lesson of the Great Famine in Greece is stark. When I asked her how a famine is stopped, she said firmly, "I think it's political will."

Nothing could underscore her assertion more than the greatest famine of the twentieth century, which ripped China apart just a little over a decade after World War II came to an end.

The Great Leap Forward That Sets Us Back

It started as a crazy dream based on the urge to transform the world. Mao Zedong, chairman of the People's Republic of China, wanted to secure his political power in the party and turn China into an industrial powerhouse that could rival Britain. He'd grown up on the utopian promises of Marxism, and as an adult revolutionary leader was awed by the massive engineering projects of the Soviet Union. So when Mao informed his fellow Communist leaders at a 1957 meeting in Moscow that China would surpass Britain in the production of basic goods like grain and steel, he drew up a plan that sounded like something out of science fiction. Under the Great Leap Forward, he said, the Chinese people would turn their prodigious energies to a massive geoengineering project—damming up some of the country's greatest rivers, halting deadly floods, and creating enough stored water to irrigate even the most arid regions in the mountains. Unfortunately, the plan turned out to be more fiction than science. Mao refused to listen to the advice of engineers, and pushed local party leaders to harness every citizen's energies to dig dams that failed and divert rivers in ways that didn't irrigate the soil. Worst of all, these projects prevented farmers from doing crucial labor on farms.

In 1958 and 1959, as Mao moved into the next phases of the Great Leap Forward, he demanded that each province meet fantastically high quotas on agricultural and steel output. According to regional records uncovered recently by the University of Hong Kong history professor Frank Dikötter, this was when the famine began to claim millions of lives—eventually killing as many as 45 million people. People died as a result of two policies: dispossession and fruitless labor. First, the government created vast work collectives by confiscating all private property, dispossessing people of their food stores, homes, and other belongings. Then local party representatives forced the understandably reluctant members of these new collectives to engage in unscientific, misguided experimental agricultural methods and steel manufacture. Farmers were told to plant rice seeds very close together, extremely deep in the soil, because it was considered a scientific method of producing a higher-yield crop. Meanwhile, to meet steel quotas and avoid punishment, people took to melting down farm equipment and anything else they could. Needless to say, the experimental farming techniques left the collectives with little food, and the steel production often left them more impoverished than ever.

Though China wasn't at war, Mao borrowed the language of militarization to propagandize on behalf of the policies that were starving his people. China's Great Leap Forward spawned terms like "the People's Army," which Mao used to characterize the displaced masses that party leaders deployed to work on China's industrialization projects. Dikötter suggests that Mao favored terms like this because being in a state of war—if only a metaphoric one—would inspire people to sacrifice even more for the good of the country. It's easy to see similarities between this strategy and the situation in occupied Greece. In both cases, war was used to justify abuses that led to millions of deaths.

China's Great Famine was the worst famine of the twentieth century, and it was entirely manufactured by human political choices, which in turn affected land use. Dikötter calls the famine a mass murder, while the Chinese government considers it to be the result of tragically misguided policies. Regardless, it's clear that the worst famines in recent history are cultural disasters rather than natural ones.

Survivors of the Great Famine included people who were willing to bend the political rules that Mao and his representatives had imposed on them. They secreted away foods that were supposed to go to the communes, engaged in illegal forms of trade, and, in a few cases, formed armed mobs and robbed trains, communes, and other villages. It wasn't until 1961 that Mao acknowledged the desperate conditions in some provinces and called off the programs of the Great Leap Forward. When people were allowed to live in more permanent homes and return to tried-and-true methods of farming, the famine slowly abated.

Are We Going to Kill Ourselves?

Stories of recent famines raise the same question that stories of war always do: Are we humans going to exterminate ourselves more efficiently than a megavolcano ever could? It's undeniable that one of the greatest threats we face is ourselves. Though famine has historically been a less efficient killer than other disasters like pandemics, and our systems for dealing with it have improved immensely over time, our survival is still at risk from malnutrition caused by environmental change and what demographer Hionidou called political will.

Evan Fraser's predictions about environmental change in North America's breadbaskets are already being borne out by the dire drought conditions that struck in the summer of 2012. Many farmers in Africa have suffered similar droughts for decades because they depend entirely on rainfall rather than irrigation systems.

Some of the environmental changes we're witnessing in the grain baskets of Africa and North America are cyclical changes that have nothing to do with humans' use of fossil fuels. But if the Intergovernmental Panel on Climate Change's recent models of rising global temperatures from carbon emissions turn out to be accurate, we'll soon be dealing with cyclical drought conditions, exacerbated by the heat humans are adding to the party. Many regions will suffer the same problems that farmers face in Africa every season, when drought can wreck an entire region's hope for food and incomes. It's very possible that our dreams for a global society in an industrialized world will have the unintended consequence of pushing most people on Earth into lives of poverty, hunger, and disease.

Leaving aside questions of environmental change, we're still contemplating an exceptionally harsh future. As UC Berkeley economics professor Brad DeLong put it to me:

> You get a famine if the price of food spikes far beyond that of some people's means. This can be because food is short, objectively. This can be because the rich have bid the resources normally used to produce food away to other uses. You also get famines even when the price of food is moderate if the incomes of large groups collapse.... In all of this, the lesson is that a properly functioning market does not seek to advance human happiness but rather to advance human wealth. What speaks in the market is money: purchasing power. If you have no money, you have no voice in the market. The market then acts as if it does not know that you exist and does not care whether you live or die.

DeLong describes a marketplace that leaves people to die—not out of malice, but out of indifference. Coupling this idea with Sen's entitlement theory, you might say oppression and war deprive people of the entitlements necessary to feed themselves. The problem is that the market doesn't care if people starve or grow ill. Based on historical evidence from famines in Ireland, Greece, and China, we can reasonably expect that if our economic systems remain unchanged, we will continue to suffer periods of mass death from famine. These famines will get worse and worse while the market continues to ignore the growing impoverished class.

Of all the forms of mass death we've looked at so far, famine can be understood as the least natural of all disasters. The good news is that famines (often accompanied by pandemics), unlike megavolcanoes and asteroid strikes, are human-made problems with human solutions. If we consider the examples of famine we've explored in this chapter, there are a few common themes that emerge in the stories of survivors. All of them have to do with ways that countries have acted collectively to fight mass death. One key lesson we can draw from Black '47 is that mobility—movement either internally or across national borders—often saves lives. A million Irish immigrants escaped death, thanks in part to other nations allowing them to relocate. Today, Somalian and Ethiopian refugees are attempting to do the same thing as they stream out of regions where food supplies have dried up. By contrast, during the Great

Leap Forward, Mao's government prevented the Chinese who lived in famine-racked provinces from fleeing to other places with better food security. The death tolls that resulted were staggering. Similarly, Greeks who suffered the worst effects of famine during World War II were trapped on islands, unable to flee even if they had wanted to risk the dangers of slipping through the blockades.

Still, global cooperation did ultimately prevent the Greek famine from reaching the proportions that the Great Leap Forward did. A few Greeks left the country, but for the most part the population was saved by humanitarian aid coming in from outside. Like immigration, food aid is a solution that requires other nations or regions to cooperatively step in. This solution to famine involves what Sen called transfer entitlements. To survive, starving regions must rely on the kindness and generosity of regions that can ship in their surplus food.

There is another lesson to be drawn from Black '47 and the Great Leap Forward that is especially important in today's drought-stricken times. Mass societies need to adapt better to their environments, figuring out ways to farm sustainably so that a few years of bumper crops don't give way to decades of blight and dust bowls. It is one of history's great tragedies that Mao's attempt to revolutionize China's land use was so horrifically misguided and ill-informed. He was right that farming methods needed to change radically to sustain China's huge population. But to say that his implementation was faulty is a gross understatement. Changes in our land use have to be based on an understanding of how ecosystems actually function over the long term. Ultimately, as we've learned from studying both human and geological history, the safest route is to maintain diversity. Farmers need to move away from specialized landscapes and monocultures that can make a region's food security vulnerable to climate change, plant diseases, and pests.

None of these solutions–immigration, aid, and transformed land use–is foolproof, certainly, but they can all prevent large groups from being extinguished. These are solutions that also require mass cooperation, often on a global scale. Preventing famine, like preventing pandemics, has meant changing our social structures. But those changes are always ongoing, often spurred by protests and political upheaval. We even have today's version of the Peasants' Revolt in the form of the Occupy movement, whose goals those London rioters in 1381 would undoubtedly have recognized and understood. Still, sometimes it feels as if change doesn't come soon enough. Famines and their accompanying pandemics are problems that we've been trying desperately to solve for hundreds of years. How are we ever going to survive over the next several hundred?

In the rest of this book, we're going to explore the answer to that question. As we've seen, human mass death is caused by a tangle of social and environmental factors. Our survival strategy will need to address both factors. We need a way forward based on rationally assessing likely threats, which we've learned about from our planet's long geological history and our experiences as a species. But we also need a plan that's based on an optimistic map of where we as a human civilization want to go in the future. To draw that map, we'll take our cues from some of the survivors around us today, human and otherwise. Those survivors and their stories are what we'll explore next.

PART III LESSONS FROM SURVIVORS

10. SCATTER: FOOTPRINTS OF THE DIASPORA

IN THE LAST two parts, we've looked at all the ways life on Earth, and especially humanity, have managed to survive hardships that ranged from meteorite strikes and megavolcanoes to the perils of migration and disease. Now we'll turn to the stories of humans and other life-forms who have survived into the present day, using techniques that could serve us well as we make plans for a future world where our descendants can thrive. We'll begin with the story of a group of humans, an ancient tribal people today called the Jews, who have retained a distinctive cultural identity for thousands of years. They've survived several deadly episodes of persecution in part by scattering and escaping in the face of adversity, rather than allowing themselves to be extinguished in the flames of war. In fact, this strategy of scattering is a crucial lesson taught to children during Passover, one of Judaism's most important cultural rituals.

When I was the youngest kid at Passover gatherings, I was given the job of reading some questions that are a crucial part of the prayers. None of them made any sense to me, including the very first one: "Why is this night different from all other nights?" As I squirmed in my seat waiting for dinner, I hoped that the answers would explain why I had to eat such weird food, like parsley dipped in salt water and sweet apples mixed with eye-watering horseradish. After many years of grumpily contemplating why I had to eat things that symbolized the tears we shed during slavery,

I figured out that Passover had nothing to do with dinner, and everything to do with memory. Passover is the one night every year when Jews retell the biblical story from Exodus about how the diaspora began. It's become such an important ritual for Jews because the allegorical stories in Exodus mirror actual catastrophic events in Jewish history. This story of survival in the Bible became, in a sense, a template for survival in the real world.

But before we consider Jewish survival in recorded history, let's recall the Passover story (apples and horseradish are optional). Thousands of years ago in ancient Egypt, the story goes, Jews lived as slaves under a cruel Pharaoh, making bricks for his pyramids and sorrowfully watching their families destroyed by backbreaking labor. Eventually a great leader came along, named Moses, and he begged the Pharaoh for his people's freedom. When Pharaoh refused, he discovered that the single, incorporeal God of the Jews—so different from his people's many half-animal gods—had some tricks up His sleeve. The God of the Jews sent ten plagues to devastate the Egyptian population, including crop-eating locusts, frogs falling from the sky, and a rain of blood (this always struck me as particularly awesome when I was a kid). In the worst of the plagues, God's "angel of death" took the firstborn son from every house that wasn't Jewish. Finally, the Pharaoh was persuaded. He told Moses to get his people out of the city, and the Jews spent one frantic day packing all their goods. They didn't even have enough time to let their bread rise, which is why we symbolize this part of the story by eating a flatbread called matzo during Passover.

Apparently, at the last minute, the Pharaoh changed his mind about the whole deal and tried to send his soldiers after the fleeing Jews. That's when Moses got superheroic, held out his hand, and parted the Red Sea. If you've ever seen Charlton Heston chewing the scenery in *The Ten Commandments,* you know what happened next. The Jews raced to the other side of the sea, hotly pursued by the Egyptian army. But once they'd reached the far shore, Moses let the waters smash back into their proper place, drowning the army and beginning the first chapter of the diaspora story.

For forty years, according to the Bible, the Jews wandered in the desert of what was then called Canaan, looking for a place to live. That's when they became a diaspora people, a group far from their ancestral home and searching for a place to live where they wouldn't be enslaved or worse. Later in the Bible, God leads the Jews to their "promised land," eventually called Israel, which their children are destined to conquer. But the story of the book of Exodus ends with the Jews still in the desert, having won one battle but facing many more, unsure whether they'll survive to find a home.

This ending is as significant as the structure of the story itself. It's oddly realistic, leaving our main characters stranded in the middle of events whose outcome only their children will ever know. It suggests that when we struggle for a better life, we may never reap the benefits of that struggle ourselves. At the same time, the meat of the story is a powerful antidote to ancient tales glorifying war that were written during the same era as Exodus probably was. Stories about how cool it is to rip your enemies' faces off appear elsewhere in the Bible (the books of Kings and Judges are complete bloodbaths), as well as in cuneiform tablets created by groups in the Assyrian empire and others. During a time in history when most nations celebrated military force and gory battles, the diaspora story in Exodus teaches us that there is great bravery in retreat. It is an act of tremendous strength to choose life and an uncertain future, rather than death in war. For the Jews who internalized this message, rather than the slaughter-is-nifty one, survival became a struggle that was often more difficult than death. But they lived. And so did their children, for generations that spanned millennia.

The First Diaspora

In modern parlance, the term "diaspora" refers to the geographical dispersion of people who are separated from their homeland. But, as political scientist William Safran explained in the first issue of the scholarly journal *Diaspora,* it can also refer to the diverse peoples who are the result of such a movement. Many groups have experienced a diaspora, including Africans outside Africa and Asians outside Asia, often due to some kind of major social upheaval. Today, these groups as well as Jews are commonly called diaspora peoples, even though many of them live in the same place that their families have for generations.

The word "diaspora" comes from ancient Greek, where it was first used to describe people who left their homelands to colonize distant regions. Gradually the term was applied to the Israelites of the era of the Babylonian exile, whose experiences were ironically the opposite of the original Greek meaning.

Though the rich geographical detail of the story in Exodus has led many to assume that it's based on an actual historical event, archaeological excavations over the past few decades suggest that the story captures the spirit of the Israelites, but not their actual historical origins. UC Berkeley archaeologist Carol Redmount studies ancient Egyptian civilizations, and says there's no evidence that the Jews or even their Asiatic ancestors were in Egypt during the time period described in Exodus—roughly during the reign of Rameses in the late second millennium BCE.

Instead, based on archaeological surveys of the region, it seems likely that the Jews during this time were a nomadic group whose members began to settle in small subsistence communities in the hills near Egypt at the height of the Bronze Age in the 1400s or 1300s BCE. Over the next several hundred years, these groups established many kingdoms, including a thriving northern region called Israel. But then in the eighth century BCE, Israel fell to the Assyrians and the formerly backwater southern kingdom of Judah rose to power. Judah's biggest city, Jerusalem, once a hick town, became a thriving, walled metropolis hugging the base of the famous Temple Mount. It was also during this period that some archaeologists believe Jewish priests in Judah put the book of Exodus together from several sources.

Still, we don't find archaeological evidence for a situation comparable to the one described in Exodus until the sixth century BCE. At that point, Judah had been a client state of Babylon for decades, and tensions between the two powers finally reached a breaking point. Judah revolted against the Babylonians and was completely crushed. In 587 BCE, the Babylonian king Nebuchadnezzar II led his troops into Jerusalem and destroyed it. Archaeologists have found sooty traces of a massive fire within the city's walls from this era, along with countless arrowheads.

The burning of the city sent many Jews into exile throughout the region, but within a few generations many returned to Jerusalem and assimilated into Babylonian society, adopting the local language, Aramaic, for writing. Indeed, in Jewish writings of the period, Judah is referred to by the Aramaic name Yehud. It's also during this era that the nomadic hill people who created the nations of Israel and Judah started calling themselves Yehudim, or Jews.

One might argue that Jewish identity coalesced during a period when its nation was fragmented. And the Babylonian exile was just the first of many great fragmentations recorded in Jewish history. In the first century CE, Jews fled the Romans; in the fifteenth century, they raced to outrun representatives of the Spanish Inquisition; and still later, they abandoned large parts of Europe to escape the twentieth-century Holocaust. Passover has probably remained such an important ritual because it's designed to remind Jews of our shared history as people who scatter in order to survive. To this day, we dwell in all the far-flung places where Jewish communities large and small continue to tell stories of a legendary time when we clung to life by running as far as we could, in as many directions as we could.

But is scattering really a good survival strategy outside of legends? If Jewish history is any guide, the answer is yes. Despite centuries of persecution and diaspora, there are people all over the world who call themselves Jews. And now we have scientific evidence that today's Jews haven't just inherited a cultural tradition. Some of us really do have biological ancestors who survived by wandering in the desert and beyond to find new homes. Population geneticists say there's strong evidence that a group of Jews originating in ancient Rome over 2,500 years ago share identifiable genetic links with Jewish populations today from Spain, Syria, North Africa, Russia, and many other places. In other words, many Jews today owe their existence to people who scattered.

The Genetic Evidence for the Diaspora

Geneticist Harry Ostrer has contributed to one of the world's largest and longest-running genetic studies of Jewish people. An energetic and talkative man, he collaborates with colleagues and subjects across the globe from a slightly cluttered office at the Albert Einstein College of Medicine, surrounded by family pictures and lab equipment. Located in a quiet neighborhood in the Bronx, the college is practically in the backyard of some of the groups Ostrer studies, like Brooklyn's Syrian Jewish community, as well as a few Iraqi Jewish enclaves in Queens. He's done work with a large group of Turkish Jews in Seattle, too.

Studying these groups and others has given Ostrer a perspective on the results of diaspora, rather than the events leading up to it. One point he emphasized strongly was that diaspora is more often about staying rather than scattering. Jewish history can be characterized by long periods of settlement and assimilation into local cultures, punctuated by sudden shifts when many people abruptly fled to new lands, usually to escape persecution. As geneticist David Goldstein notes in his book *Jacob's Legacy: A Genetic View of Jewish History*, some of the earliest historical records of the Jews come from sixth-century BCE cuneiform tablets, which describe the Babylonian conquest of Jerusalem. But the results of that diaspora are lost to history. The next great Jewish settlement took place in Rome, and the diaspora that resulted has genetic echoes all the way up into the twenty-first century. The Roman diaspora is the focus of Ostrer's research.

There are extensive records of Jewish culture from the Roman empire during the first century CE. Many of these Jews were brought to Rome as slaves starting in the second century BCE, from Greece, Judea (the former southern kingdom of Judah), and many areas in the region. Over the next century, Jews assimilated into Roman culture and became one of the biggest and most powerful minority groups in the empire. Though we have no reliable source for how many Jews there were in Rome, we know from contemporary sources that Judaism was a highly visible religion. Politicians issued laws regulating the practice of Judaism, and many Jews became Roman citizens. Meanwhile, in the courts, commentators often complained of Jewish "disturbances"—probably referring to political unrest in response to constantly shifting Roman rules about Jewish taxation and social status.

Unlike today, Roman Jews expanded the ranks of their temples by actively proselytizing. They assimilated into Roman life, but Romans assimilated into Jewish traditions, too. It was a time of great cultural mixing that finally came to an end in the late first century CE, when Emperor Claudius ordered all Jews to be expelled from Rome. A few years later, some Jews in Jerusalem rebelled against the Roman control of their city and were defeated, while Roman Jews fled their homes to avoid death or worse. In the Bible, this period is referred to as the time of the Second Temple's destruction because the Romans destroyed the house of worship on the Temple Mount just as the Babylonians had over 600 years before.

Though this diaspora survives in historical documents and biblical stories, Ostrer wanted to know if he could track down evidence of a direct genetic connection between the Jews who left ancient Rome and the Jews alive today. To find out, he had to get DNA samples from hundreds of Jews across the world, looking for genetic commonalities. "I went to Rome and did recruitment there," Ostrer recalled. "That's been a stable community for hundreds of years and perhaps dates back to the community that was there in classical antiquity." He also got samples from Eastern European Jews as well as Jews in immigrant communities in the New York area. Anybody who could trace their Jewish ancestry back two generations to all four of their grandparents was eligible to participate.

Once he'd amassed his samples, Ostrer and his team had the beginnings of what they call the Jewish HapMap. "Hap" is short for "haplotype," a term geneticists use to describe a set of unique genetic markers in the human genome. People who share haplotypes are more closely related to one another than people who don't, and Ostrer wanted to know whether he could identify distinctly Jewish haplotypes. Over several years, the researchers at the Jewish HapMap Project scoured their data using a variety of statistical methods to compare both short and long strands of DNA from volunteers. They began to see patterns suggesting that people who had lived close together centuries ago still shared genetic similarities. Jews in Central Europe today share more genetically with Jews in the Middle East than a non-Jewish person living in Central Europe does with a non-Jewish person in the Middle East. And it's all because those groups of contemporary Jews had ancestors from the same regions of Rome. Discoveries like this demonstrated that there are distinctive Jewish haplotypes that offer hints about where people's ancestors settled in the diaspora.

Once they had enough data, Ostrer and his colleagues could actually create genetic maps tracking the spread of Jewish haplotypes out of ancient Rome and into the Middle East and Europe. Why were they able to isolate these haplotypes at all, when so much time had passed? It had to do with a change in Jewish culture after the Roman diaspora. Jews in ancient Rome were proselytizers—they converted many people and intermarried with non-Jews regularly. The Jews of that era would probably have shared haplotypes with their Jupiter-worshipping neighbors. But after their expulsion from Rome and the destruction of Jerusalem in the first century CE, Jews changed the structure of their communities radically. No longer were they permitted to proselytize and intermarry. To be considered truly Jewish, a child had to be born of a Jewish mother, establishing a rigorous matrilineal line. Without realizing it, the Jews of the first century created a culture that allowed their unique haplotypes to endure over the next 2,000 years.

Mapping the diaspora becomes more difficult when you add in the evidence of extensive assimilation and intermarriage taking place in Europe before the Inquisition. In countries like Spain, Jews enjoyed a social status comparable to the one they had once held in ancient Rome. They were prominent members of their cities, intermarried with non-Jews, and dramatically expanded their communities. But the tide turned in the fourteenth century, which saw the rise of political persecution of Spanish Jews. This culminated in the fifteenth century as the Spanish Inquisition spread outward into Portugal and Rome, and once again sent Jews running into their familiar diaspora pattern, pushing them deeper into Europe and the East. Still, they survived and even retained some of their haplotypic particularities. A group of Portuguese anthropologists recently discovered a small group of Jews living in the mountains of Portugal whose ancestors had apparently fled there and masqueraded as Catholics to escape the Inquisition.

Despite what he and other geneticists have discovered, Ostrer is wary of saying too much about the genetic basis for Jewish identity. This is an area of inquiry that is still evolving rapidly, and he's quite willing to admit that some of his conclusions are simply "a guesstimate." There is no single haplotype that unites all Jews–instead, he and his team found four distinct haplotypes identified with different Jewish diaspora groups. There will never be a genetic "Jew or Not" test. All that Ostrer's work reveals is that a genetically identifiable "Jewish people" survived the diaspora. We now have both historical and genetic evidence that scattering and hiding out during times of upheaval is a good way to ensure that your progeny will survive–even for dozens of generations.

The Black Atlantic

Toward the end of my conversation with Ostrer, we started talking about Jews today. We're in the midst of another period of Jewish assimilation and migration, he said, making a sweeping gesture with his hands as if to encompass all of New York, or possibly the world. In the wake of nineteenth-century pogroms and the twentieth-century Holocaust, many Jews were forced to scatter to new areas. And some, like Reform Jews in the United States, have started converting people to Judaism again. The result of all this movement and intermixing is a Jew like me. My mother was a Methodist who converted to Judaism before she married my Jewish father. I was raised Jewish, but who knows what kind of haplotype I have? More to the point, when we're talking about the survival of a group over centuries, does it really matter whether I'm culturally Jewish or genetically Jewish or somewhere in between? After hundreds of years of diaspora, aren't all survivors a little bit hybrid?

This is exactly the question that people from many diaspora groups have raised over the past half century. Perhaps nowhere is the answer to it more beautifully expressed than in the book *The Black Atlantic: Modernity and Double-Consciousness*, by Guyanese-British scholar Paul Gilroy. While researching the often fragmented histories of blacks in England, Gilroy realized that he should reframe black identity as a hybrid experience that combines many cultures. To describe the origin of this experience, he called on the idea of a "black Atlantic," the geographical region where African slaves were scattered in a forced diaspora across Europe and the Americas. Instead of having a single point of origin, like the lands around Jerusalem, Gilroy's diaspora has many origins. And its survivors are genetic and cultural hybrids. But that doesn't mean African identities have been extinguished in people outside Africa today. It has survived in a multitude of ways, though some of them might be unrecognizable to communities who lived in Africa half a millennium ago.

As Ostrer put it, diaspora is about where you come from, but it is also about where you end up. Our journeys change us radically, but when we settle down again there is a continuity, a shared history that holds us together. Jews and Africans are not unique in this respect—many groups have maintained a sense of community through times of hardship and separation. Recent human history teaches us that your group has a better chance of surviving in the long term if you're willing to divide into groups and go your separate ways to safety. But that doesn't mean the past is lost. What makes a book like *The Black Atlantic* so important is Gilroy's powerful assertion that even if your group is unwillingly torn apart and assimilated into other cultures, your progeny will remember where they came from even hundreds of years from now.

The Passover ritual makes a similar assertion. It is a celebration of identity forged in diaspora, and a reminder that survival often means finding a new home. The difficult part, as we face an uncertain future, is how to understand the meaning of "a new home." We may have to look very far afield to get our answers. In fact, one of the greatest stories of survival through adaptation does not come from humans at all. It comes from the humble blue-green algae, whose incredible history may also show us one possible path into the future.

11. ADAPT: MEET THE TOUGHEST MICROBES IN THE WORLD

YOU'VE PROBABLY CALLED it scum. But that slimy blue-green goo floating in ponds and on the ocean comes from a group of species so hardy that humanity's fumbling attempts to adapt to our environments would be a joke to them. Well, it would be if these blobs of raw biological productivity had a mean sense of humor, or brains, or even mouths to laugh with. We're talking about our old friend cyanobacteria, whom we met billions of years ago, in the first chapter of this book. At that time, it was busily unleashing enough oxygen to transform the composition of Earth's atmosphere. Its subsequent 3.5-billion-year career as a life-form proves that this ancient breed of scum has gotten something fundamentally right. Cyano, as it is fondly known among scientists, evolved one of the planet's greatest adaptations: photosynthesis, or the ability to convert light and water into chemical energy, releasing oxygen in the process.

Cyano has also had a secondary career as a biological building block for other life-forms. About 600 million years ago, sometime before the first multicellular life appeared, cyano began forming symbiotic relationships with other organisms, slowly merging with them over the millennia. Eventually these early cyano evolved inside other cells to become chloroplasts, tiny organs (known as organelles) that handle photosynthesis for plant cells. Every plant on Earth is, in fact, the result of this merging process. You can think of chloroplasts as both engines and batteries for plant cells; photosynthesis creates forms of energy that plants can use immediately as well as store for later. Cyano's great adaptation is so powerful that plants and even a few animals like sea anemones have survived by absorbing cyano and turning it into their own adaptation.

Brett Neilan, a biologist at the University of New South Wales, has spent his life studying cyano among the ancient rocks of Australia's coastline, and he thinks the secret to the algae's success is simple. Cyano's ancestors won the evolution game because they worked with what the Earth always had in good supply: sunshine, or some form of light, and water. Like most plants, cyano are called autotrophs, a word that means "self-feeding," and refers to their ability to feed themselves without consuming other organisms. In a sense, cyano generate the food they consume. As a result, they can and do live everywhere. They've been found in Antarctica and in the boiling, acidic waters of Yellowstone's geysers. Not only are they seemingly impervious to dramatic temperature changes, but they are virtually immune to famine as well. According to Neilan, cyano prevent themselves from starving in times of scarcity by storing extra nutrients like nitrogen in little sacs tucked inside their cellular walls. If these food caches are not enough, cyano can go into stasis. The microbes put themselves into a kind of suspended animation and can endure without food for years, waiting out droughts or other disasters that affect their food supply.

Cyano have other incredible abilities, too. They can live as individual single-celled organisms, but they can also join together with other cyano, Mighty Morphin Power Rangers–style, to form a multicellular creature. They are the simplest organism on Earth whose biological processes are regulated by the circadian rhythms of light and dark. Like humans, with our sleeping and waking cycles, cyanobacteria engage in different metabolic activities depending on whether it's day or night. This allows them to engage in two separate chemical processes for nourishment–photosynthesis and nitrogen fixation–which would normally interfere with each other. Thanks to their circadian clocks, cyano can do photosynthesis by day and nitrogen fixation by night. They thus benefit from two kinds of nutrient production. Other plants benefit, too. Just as some organisms absorbed cyano to create choloroplasts, others have formed symbiotic relationships with the bacteria to reap the benefits of the energy produced by nitrogen fixation.

Cyano have succeeded so well on Earth because they create their own food, using a power source that is ubiquitous and sustainable. It's such a good strategy that other life-forms learned from cyano's success millions of years ago and absorbed these tiny engines into their own fuel-production processes. Humans may not be able to merge with cyano on a biological level—at least, not with current levels of technology—but many scientists are working on ways we could use photosynthesis to create more sustainable energy sources to help humans survive as a mass society.

Why Is Photosynthesis So Awesome?

One of these scientists is physicist-turned-biologist Himadri Pakrasi, who runs Washington University's International Center for Advanced Renewable Energy and Sustainability (I-CARES). With a thatch of curly black hair just beginning to turn gray and a ready smile, Pakrasi radiates enthusiasm for his work. The first time I spoke to him, by phone, it was to find out how his lab had managed to create energy using water, light, and bacteria. "You should come out here and see!" he exclaimed. Very few scientists would invite a writer they'd never met before to visit their labs, but Pakrasi is the kind of guy who wants to get people engaged with his work—even strangers from halfway across the country. It was easy for me to understand how he'd built up a large international group of collaborators at Washington University, including scientists, city planners, and engineers.

When I arrived in St. Louis a couple of months later, Pakrasi told me that he'd been fascinated by photosynthesis his whole life. "Every plant is a fantastic power reactor," he explained. "Let's learn from nature how to do that ourselves. Let's have a perpetual synthetic plant that makes energy." He and his colleagues at I-CARES are convinced humans could be using algae to fuel our cities in a century. The cornfields outside Pakrasi's office window would bloom with photosynthetic antennae, or super-efficient solar cells atop flexible structures, their light-consuming faces twisting to follow the path of the sun across the sky. Energy breweries the size of local beer megacorp Anheuser-Busch would be packed with vats full of bubbling blue-green algae that could be used in batteries or other chemical processes. Humanity would survive the fossil-fuel age by drawing energy from cyano. But before Pakrasi's visions can come to pass, scientists need to figure out how photosynthesis works.

In his lab at Washington University in St. Louis, researcher Himadri Pakrasi shows another researcher some of the cyanobacteria colonies that have been engineered to produce higher amounts of hydrogen.

Despite what you may have learned in high school biology, photosynthesis isn't simple. In fact, it's a chemical process that follows some seriously weird and mysterious pathways—some of which we still don't understand. Another Washington University professor, the physicist Cynthia Lo, flipped her laptop open to show me her work on photosynthesis, glanced at some diagrams, and looked momentarily exasperated. "You know why most plants are green?" she asked rhetorically. "It's because they're terrible at capturing and absorbing green light. So they capture blue light, but they reflect green. And that's what you're seeing in this bright green algae." Lo is one of Pakrasi's research collaborators at I-CARES, and the principal investigator on the Photosynthetic Antenna Project. She's working out the basic science that might one day lead to Pakrasi's vision of superefficient solar cells collecting light to power the city of St. Louis. Lo clicked through some diagrams of how photosynthesis works at the atomic level, photons colliding with molecules called pigments to produce energy.

Then Lo returned to a theme that would come up a lot in our conversation: cyano are actually terrible at reaping the benefits of photosynthesis. Not only are they missing out on green light, but they only convert about 3 percent of the light they harvest into energy. By comparison, commercially available solar cells convert about 10 to 20 percent of incoming light into electricity. But, Lo said, today's solar cells can only harvest a small percentage of the light wavelengths that cyano collect—so the bacteria are still way ahead of us in that department. But not for long, if Lo and her lab have anything to say about it.

Lo's research into the physics behind light capture could help engineers build solar cells that replicate the molecular smashup we see during photosynthesis. Engineers call this biomimesis, or the practice of imitating biological forms to make artificial systems work as efficiently as living systems do—or more efficiently. "A biological system is intriguing because nature has optimized it," Lo explained. But it's not optimized enough. Algae harvests light really efficiently, but doesn't convert it into energy efficiently. Solar cells are efficient at making energy but not at light harvesting. Ultimately, Lo's goal is to figure out what it would take to develop what she calls a biohybrid solar cell that combines the light-capturing abilities of cyano with the energy-conversion abilities of existing solar-energy technology.

By trying to copy the energy reactors inside each cyano cell, Lo and her team are learning the best possible lesson they can from this mega survivor. They are trying to diversify our energy supply, creating new ways for us to gain energy from the environment so that we can survive long-term with a sustainable electrical grid. It may be decades before we crack the code on photosynthesis, but this ancient organism could guarantee a better future for the planet—just the way it did billions of years ago.

Turning Coal Plants into Cyano Breweries

Another of Pakrasi's collaborators is working on a strategy to take us from a world run by coal to one powered by plants. Environmental engineer Richard Axelbaum, a wiry man whose office desk is decorated with angular chunks of coal, is interested in the near future of alternative energy. Pakrasi and Lo are looking perhaps half a century ahead, while Axelbaum looks just 10 to 20 years out. He has to be a pragmatist. That's why he works on "cleaner coal" technology and carbon sequestration, the practice of sustainably disposing of coal's greenhouse gas by-products.

One of his projects is a prototype coal-combustion facility called the Advanced Coal and Energy Research Facility, located in a huge, high-ceilinged warehouse on the Washington University campus. The facility sustains tanks of healthy algae using a by-product of coal processing. From a viewing gallery two floors above, Axelbaum showed me a tangle of thick pipes, cylindrical tanks, and a grid of shelves packed full of bubbling aquariums. Axelbaum pointed to a tank that looks like an outsized metal barrel turned on its side. "That's the coal-combustion chamber," he explained. Unlike typical coal-burning plants, this chamber burns the coal in a pure oxygen environment. As a result, the only by-products of the process are "cleaner" because they're composed almost entirely of carbon dioxide and ash, with no nitrogen compounds mixed in. "Every generation has had its clean coal," Axelbaum remarked. Early twentieth-century facilities improved on the extremely dirty coal-burning practices of the nineteenth century, for example. And now he's hoping that we can improve the process even more, bringing us one step closer to truly clean energy.

Axelbaum's finger followed a thick duct emerging from the combustion chamber. "That goes to a white-ash capture chamber," he said, identifying a big, rectangular bin. Normally, coal ash is stored in large open-air ponds, which can cause environmental damage. "Our hope is that all this ash can be put to use, whether in concrete or new kinds of conductive materials," Axelbaum said. As for the carbon dioxide? "That's going over to the algae tanks." Axelbaum pointed at pipes leading to the aquariums. The algae absorb the carbon, thriving on the gas. Axelbaum's oxy-coal combustion could be feeding (literally) the next generation of superclean energy production.

The Algae Economy

A couple of years before I visited Pakrasi, his team made an incredible breakthrough. They were working with a mutant strain of cyano that releases hydrogen instead of oxygen during photosynthesis, and they managed to coax the algae to produce ten times more hydrogen than other strains had. Hydrogen is often called a clean fuel because when it's burned it releases mostly water. Hydrogen fuel has been used for rockets, but its production is too expensive for consumer markets. Still, its widespread use in every home is part of the future of cyano-powered energy that Pakrasi, Lo, and Axelbaum dream about.

Imagine a world where brewers grow hydrogen fuel by feeding cyano with the carbon dioxide released from burning coal. The Pakrasi lab's cyano also consumes glycogen, a by-product of biodiesel production. So basically, these algae cells are eating two harmful by-products of energy production to produce a form of fuel whose consumption releases almost no toxins at all. "They give you a lot of bang for your buck," Pakrasi said with a laugh. Eventually, we could wean ourselves off coal and make the leap into a cyano-powered world full of new kinds of green fuel.

Pakrasi imagines a future where biologists could even develop specific strains of cyano to transform all aspects of industrial production. The bacteria could eventually replace petroleum, and aid in the production of chemicals like polypropylene, which is used in the synthesis of everything from rope and lab equipment to thermal underwear and durable plastic-food containers. Famed scientist and U.S. secretary of energy Steven Chu has talked about replacing the oil economy with a biofuel "glucose economy." But Pakrasi and his colleagues in I-CARES have refined this notion even further, and speculate about a global algae economy whose engines run on photosynthesis.

Pakrasi, who studied physics in India before coming to the States for his Ph.D. in biology, says he often looks to India and China for inspiration when he thinks about how to implement the discoveries he's making in the lab. "It's hard to [test new energy systems] here or in Europe because these countries have stable infrastructures that are already built. We're always trying to catch up, to retrofit," he mused. "But in China or India, it seems like every millisecond they are setting up new structures. These are the places where the technology we're developing here can be applied directly." Under Pakrasi's guidance, I-CARES has developed strong relationships with universities in India and China, and researchers in St. Louis collaborate with colleagues across the world. They're even reaching out beyond the sciences, to bring in experts in ethics and sociology. "As scientists, we're good at coming up with technical solutions," Pakrasi said, "but as far as the policy and human angles, we have to collaborate with [other branches of the university too.]"

I-CARES is the kind of institution that we'll be seeing more often at universities and in industry, combining people from many disciplines to come up with global solutions to problems that straddle the line between science and society. Already, the U.S. Department of Energy has funded a massive effort in California, the Joint Center for Artificial Photosynthesis, whose aims are similar to I-CARES. Its team of over a hundred scientists, many based at Caltech and the Lawrence Berkeley National Laboratory, aims to develop a way to extract clean energy from sunlight, water, and carbon, just the way plants do.

This futuristic collaborative research could one day save the world. And it grew out of the simple cyanobacteria and its best lesson, which is to adapt and diversify by taking advantage of a sustainable form of energy. In the next chapter, we'll learn about another life-form with an extraordinary survival mechanism—one that may have helped bring it back from the brink of extinction. You might say that this animal, the gray whale, lives by memory alone.

12. REMEMBER: SWIM SOUTH

GRAY WHALES JUST look like survivors. Their slate-colored skin is crusted with barnacles, and their huge, scarred jaws curve downward in what seem to be permanent grimaces. Bottom-feeders who mostly eat tiny crustaceans, these creatures nevertheless have a reputation as formidable fighters. Only packs of orcas and humans usually dare to hunt them, and accounts going back several centuries describe the deadly wrath of grays pursued by whalers. In 1874, the whaler and naturalist Charles Melville Scammon wrote about his experiences hunting grays. He recalled, "Hardly a day passes but there is upsetting or staving of boats, the crews receiving bruises, cuts, and, in many instances, having limbs broken; and repeated accidents have happened in which men have been instantly killed, or received mortal injury." Grays, he explained, possessed "unusual sagacity," which made them a hard target—especially when the animals' intelligence was coupled with their 35-to-50-foot lengths, 80,000-pound bodies, and "quick and deviating movements."

Despite their ferocity, grays have one vulnerability. Every winter, they migrate thousands of kilometers from the safety of their Arctic Ocean feeding grounds to a series of warm lagoons in Baja California, Mexico. One of the most popular spots is nicknamed Scammon's Lagoon, after the whaler. Theirs is close to the longest migration taken by any animal

The gray whale travels on one of the longest migrations of any animal in the world.

on the planet, and the whales will encounter many predators and treacherous conditions along the way. Then, after a winter spent having children (and making them) in the lagoons, they begin the trip back up the coast again, often tailed by their young. Though both the Arctic Ocean and Mexican lagoons are relatively sheltered from predators by natural barriers, the long migrations in between leave the whales exposed to danger for months at a time. How do they manage?

Grays have evolved a number of features that seem to protect them during their migrations. Remarkably, the whales never stop swimming during these journeys. Their brains "sleep" by shutting down only one hemisphere at a time, so one part of the gray's brain is always awake to keep it moving in the right direction. Even more unbelievably, grays rarely pause to feed during their migration. Instead they live on stored energy. They've spent the entire summer grazing on the Arctic seafloor, building up a thick layer of energy-storing blubber which they burn through during the roughly seven-month round-trip to Mexico. Grays eat by taking giant bites of dirt and sifting tasty crustaceans out through the baleen filters in their mouths. This is why they're often seen with big, muddy smears on their lips after they eat. Marine biologists often jokingly call them the cows of the sea. Grays spend half the year eating so that they can spend the other half migrating and reproducing.

It's likely that grays have been living this way for the many millennia since they first evolved 2.5 million years ago. Grays are also slightly less complex than some other cetaceans, which has led some biologists to speculate that they are a more ancient species. They don't "sing" by creating complex harmonies like humpback whales do. They emit what scientists call moaning noises that can be heard only at close range—unlike humpback songs, which can be heard for kilometers underwater. Though grays are able learners, as Scammon observed over a century ago, they don't exhibit a lot of social behavior like their cetacean cousins the dolphins. Instead of swimming in pods, they prefer to migrate in loose, ever-changing groups of two or three. Many travel alone. Still, grays have maintained what could be called a tradition, their great migration, that gets passed from one generation to the next. This isn't a matter of mere instinct. Scientists believe it's something that each new generation of juveniles must learn from the adults, like passing along a map that is vital to the survival of the species. It's therefore no exaggeration to say that grays survive by relying on their memories. Without memory, they would never find food, nor enjoy a mating season.

Humans nearly drove gray whales to extinction in the early twentieth century, but thanks to one of the earliest conservation agreements in the world, the gray population today has rebounded to what it may have been before whalers thinned the animals' ranks. The story of gray whale survival offers us two lessons. It teaches us the importance of passing along knowledge from one generation to the next, and it shows us one sure way to stop extinction in its tracks.

Migrations and Memory

People have been observing gray whales for centuries, but there are still many aspects of these creatures' lives that remain a mystery. Often, we only catch glimpses of their behavior when the whales are in trouble, straying from their usual paths. This was certainly the case in 1988, when an Inuit whaler spotted a group of three grays stranded in the Arctic waters. It was so late in the season that ice had blocked their path out to the northern Pacific. Grays begin their southern migration when the Arctic starts to freeze. If they stay to graze a little too long, they get boxed in by ice that's formed over the top of the ocean. With no room for the animals to surface and breathe, the straggler grays drown. It happens to a few whales every year, and locals are used to seeing their bodies wash ashore after the ice retreats in summer. But these grays hadn't drowned yet–in fact, all three (including a small calf) were surfacing to breathe out of a small open hole in the ice. Footage of their struggle to survive captured national attention, bringing television crews and scientists flocking to the small Alaska town where the creatures were stranded.

A young biologist named Jim Harvey came too, trying to reconcile the behavior of these grays with what he'd seen before. These three were clearly working together to share the airhole and survive, though typically grays are solitary creatures. What's more, the grays seemed to figure out that the humans jumping up and down on the ice around their hole wanted to help them. Eventually, after forces from both the Soviet Union and the United States got involved in the quest to free the whales, the grays followed an icebreaker out to the open sea. Harvey, now a professor at Moss Landing Marine Laboratories (MLML) on Monterey Bay, has spent the decades since the incident studying marine mammals and other creatures that make a home on the shoreline.

When I visited Harvey at MLML, a cluster of artfully designed, recycled wood buildings built just a few yards from the waters of the bay, the door and windows in his office were thrown open. Outside, seabirds skimmed over the sunny water, and grass furred the sand dunes. Further out to sea, sea lions barked and frolicked in waters where the grays travel twice a year. For decades, Monterey Bay has been a prime spot for gray whale observation–it seems to be a favorite place for the whales. Here they swim very close to shore, making it easy to take population counts and watch them in the wild.

From decades of observation, it's become clear that the whales don't choose just one group of companions for the whole migration. "They'll be with a bunch of animals, forming and changing groups all the time," Harvey told me. "It's like being in a bicycle race. You can draft behind [the leader], and it's nice to be in a group because the guy in front is usually paying attention. I think gray whales do that, too. They trade positions in terms of paying attention." Harvey had just come in from a run along the water, where he'd followed a narrow trail between MLML, a few other local marine-biology labs, and the undeveloped coastline.

His mind still on the dynamics of racing, Harvey pondered a question that is hotly contested among biologists. How, exactly, do the grays learn to navigate their way along all those thousands of kilometers of coastline? "I'm purely speculating," he said, "but I think they're following each other, and somebody else follows them, and they remember it." When I asked whether they're communicating directions with sound, too, he shook his head. "I'm sure they don't talk to each other. They're just following each other." Young whales always make the trip with an animal that has gone before.

Still, the trip changes year by year; grays are constantly tweaking their route. Twenty years ago, most of the grays migrated along a path that took them inside the Channel Islands, and closer to the coastal cities of Santa Barbara and Los Angeles. The problem was that they stuck to the shoreline too closely, often following it all the way into the shallow waters where they would become trapped. Grays have had similar problems getting lost in San Francisco Bay and Monterey Bay when they chart their course using what Harvey jokingly referred to as the "keep the shoreline on the left" method. But today, their routes take them outside the Channel Islands, and often outside San Francisco Bay too. "So they've figured it out," Harvey said. They realized that more direct routes away from the coast would be faster and less dangerous, and passed that information on. Grays live for about 50 to 70 years, so these course corrections are taking place within the lifespan of a typical animal.

More recently, Harvey and a National Oceanic and Atmospheric Administration (NOAA) biologist named Wayne Perryman have observed that grays are migrating later in the year, possibly because melting Arctic ice means they have to go farther north to find good grazing grounds. As a result, a faster route south is going to become more desirable to the grays as the years go by—they need to cut corners, as it were. But this longer route is also changing a lot more than their maps to the south. In 2012, observers were surprised to find a female gray and her very young calf swimming in San Francisco Bay. Given the age of the calf, Harvey and Perryman estimated that the gray had probably given birth en route to Mexico. She'd left the Arctic so late in the season that she wasn't able to get there in time to have her baby. It's possible that the melting Arctic ice will dramatically change the migratory cycles of the Pacific grays, altering the map that one generation of whales passes along to the next. This is another clue that the grays navigate their migratory routes by learning and memory—if the trip were somehow hardwired into their brains, they wouldn't be able to shift its parameters every year depending on environmental conditions.

Of course, some grays don't manage to remember the route quite right—which is why, for example, those three grays got caught in the frozen Arctic in 1988 and had to be rescued by icebreakers. This leads Harvey to another big question. Why should the grays continue to migrate at all, as the thawing Arctic slowly becomes more habitable year-round? "It might get to a point when they don't have to go, but the reality is that the water is still cold," he mused. And staying warm in winter Arctic waters takes a lot of energy. He and his colleagues believe it's worth it for the whales to swim all the way down to Mexico and save energy in the warm water, rather than not swimming but remaining in the cold water.

This also helps to explain why juvenile whales make the journey down to Mexico, even though they are still too young to participate in the mating and calving that goes on in the lagoons. To get the most out of all the blubber they've been building up in summer, the young whales need to seek out warmer waters with their elders. But there's another benefit, too. "Eventually, if you want to be part of the reproductive group, you need to know how to do the migration," Harvey explained. "They are gaining knowledge, including reproductive knowledge, by making the journey."

It's likely that the young whales are learning another survival skill along with the route south and then north again. When they arrive in Mexico they're watching other grays reproduce. How whales learn to mate is a big question mark scientifically, but Harvey said it's possible that they do it the same way they learn to migrate: through observation and memory.

How the Gray Whale Came Back from Extinction

Unfortunately, memory is no defense against the concerted efforts of ships full of people with harpoons and explosives. The descendants of the whales Scammon hunted still roam the waters of the Pacific coast, but their now-extinct relatives in the Atlantic weren't as lucky. A large group of grays lived in the Atlantic for thousands of years, migrating from the Arctic to the Mediterranean. But historical evidence suggests that they succumbed to hunters in the eighteenth century. Today, there are only two groups of gray whales left. One, the eastern Pacific, or California-Chukchi group, whose migration we've talked about up to this point, contains perhaps 20,000–30,000 individuals. The other is a small, poorly understood group of roughly 200 individuals called the western Pacific or Korean-Okhotsk grays. These whales have a different migration route, along the coast of Asia. In summer they graze along coastlines in the Sea of Okhotsk off the coast of Russia, above Korea and Japan. Their calving grounds are off the coast of Korea.

Both these groups would have gone the way of their Atlantic cousins if it hadn't been for the rise of conservation groups in California during the early twentieth century. After the establishment of groups like the Sierra Club, which helped protect Yosemite National Park from development in the late nineteenth century, the burgeoning environmentalist movement began to think about protecting animals as well as environments. Even whalers like Scammon noted with unhappiness that the whales were going to be exterminated if hunting kept up at the pace he observed. Often, hunters would simply plow into the mating lagoons and slaughter the vulnerable mothers and calves, destroying the population's ability to reproduce. Disturbed by the inhumanity of these hunting practices, and aware that grays weren't particularly valuable as commodities, in 1949 the newly formed International Whaling Commission outlawed the hunting of gray whales. Since that time, many scientists believe that the eastern Pacific population has bounced back to what it might have been before whaling started. Others argue, based on genetic data, that it's likely the original population before whaling was closer to 90,000 individuals.

Regardless of what the original population was, marine biologists who study the whales seem to agree that the California grays have rebounded in an extraordinary fashion. Over the past 60 years, they've gone from near extinction to a healthy, diverse group capable of learning new migratory strategies to cope with changing conditions in the Arctic. Their growing population numbers stand in stark contrast to those of other whales, especially ones like the right whale, humpback, and blue whale that navigate their own great migrations every year. Several studies suggest that noise pollution in the water from radar, and human encroachment into their territories, may be disorienting these whales. This leaves them vulnerable to beaching, or injuring themselves by swimming straight into large ships.

Grays were saved from extinction because humans chose to change their behavior. It's possible that humans might save other whales from extinction, too, by changing our behavior in the same way. We could avoid using frequencies that whales prefer for their sonar. Or we could track whale migratory patterns using satellites—something that many scientists do already—and avoid creating shipping lanes near the areas where whales are making their journeys. Changing the way we use sonar is obviously more difficult than outlawing whale hunting. But it is certainly possible, and today's gray whale population is a reminder that extinction is not inevitable for these massive sea mammals.

There's another reason grays are good survivors, though. Their migratory patterns keep them relatively safe and well fed. Few animals compete with them for food in their Arctic Ocean hunting grounds, which are vast and well stocked. Humpbacks, by contrast, feed in a small coastal area, which Harvey calls "a very compressed region." If they can't find prey in that region, humpbacks suffer. But the grays have found an enormous feeding ground where they can chomp on the seafloor during summer, as well as a protected place to mate in winter. And they've carefully charted a route between the two places that's as safe as possible. Traveling along the coasts, their large bodies are generally hidden from predators by the sounds of the surf and the dirty, silty water they love. Their lack of sonar may mean that grays are more solitary, simple creatures than some other cetaceans. But this simplicity has helped them rebound from extinction, while whales with more complex communication and social structures are suffering.

Still, the grays would never have made it this far without their ability to pass along a survival map from one generation to the next. As long as they keep learning new ways to survive the arduous Pacific coastal migration, the grays will endure.

Always Coming Home

Nomadic humans survived for thousands of years in a similar way, wandering across vast regions to find food and good seasonal weather. Many human tribal groups had traditions where they met once or twice each year for large gatherings, not unlike what the grays do when they converge in the Mexican lagoons. At these gatherings, nomadic humans would exchange gifts, pass on stories, and find people to marry outside their bands. Today, however, human survival can't hinge on migratory patterns. Most humans live in settled communities and cities, and the knowledge we pass on to the next generation is infinitely more complex than a migratory route or information about where to find the most abundant food. We've learned so much that we need libraries and databases to augment our memories.

Still, the grays have a lesson to teach us about the role of memory in survival. This struck me forcefully one afternoon in Monterey Bay, when I joined a whale-watching group in a small boat that fought its way over wind-whipped waves in search of the elusive migrating grays. The most adventurous of us made our way to the bow with our cameras, clinging to the railings and getting completely soaked by spray. We were joined by several large schools of porpoises. A group of four kept jumping out of the waves at the same time in graceful synchrony, as if trying to make it clear to the ridiculous monkeys who really belonged out here. But we kept scanning the horizon for grays, hoping to see their characteristic spouts. At last, just when we were about to retreat, we spotted one. The whale slid its blowhole just above the waves, most of its great bulk obscured by water and distance, and disappeared again. We all jumped up and down, pointing and forgetting to take pictures in our excitement. Just the sight of such a magnificent creature filled us with crazy awe, and all the sopping people in the bow started bonding and swapping stories of other amazing animals we'd seen. One person had seen a Bengal tiger in the wild, and another had been in Mexico to see the grays in their winter home. Nobody forgets these kinds of sightings because most of us, no matter how much we love urban life and civilization, also deeply love nature.

Like grays, humans are good survivors because we've learned to find food and homes across a vast region of the planet. Also like grays, we've learned to traverse these territories in more efficient ways, responding to changes in the environment. We've figured out how to build cities that protect us better than villages did; we've passed along stories of how to survive best on what is still a dangerous planet. Sometimes, we've even changed our behavior to protect life-forms other than ourselves. As we turn to the next part of this book, about planning for the future, we're going to remember the grays' lesson: You're always coming home, but the path to get there is going to change all the time.

13. PRAGMATIC OPTIMISM, OR STORIES OF SURVIVAL

IN THE PREVIOUS three chapters, we've zeroed in on strategies that have helped three different life-forms—humans, cyanobacteria, and gray whales—survive in extremely adverse conditions. We learned how an ancient tribe of humans, today called Jews, lived by scattering and founding new communities in the face of war and oppression. We explored how cyanobacteria's ability to generate its own form of sustainable energy has made it perhaps the most adaptable life-form on Earth. And we followed a group of gray whales on their difficult migration down the eastern Pacific coast, a journey each whale has memorized in order to survive, and even to bounce back from extinction once humans agreed to stop hunting them. By passing along stories about these survivors, we learn what it would take for humans to survive, too. But some stories about survival are more helpful than others.

In part two, we explored the role symbolic communication played in human evolution. Storytelling could be called the cultural backbone of human survival. There's a reason that conquering armies often burn the books and libraries of their enemies. Extinguishing a people's stories is a way of erasing their future. But when we remember those stories, they can steer us in a direction that leads away from death. In fact, stories about how humans might live in the future—sometimes known as science fiction—may be among the most important survival tools we have. We can use these stories as a highly symbolic version of the migration maps that gray whales pass on to the next generation. Futuristic stories offer possible pathways our species can take if we want our progeny to thrive for at least another million years.

What Makes Us Want to Survive?

One of the twentieth century's greatest science-fiction writers, Octavia Butler, told *Essence* magazine, "To try to foretell the future without studying history is like trying to learn to read without bothering to learn the alphabet." Butler grew up during the height of the space race in the 1950s, surrounded by hopeful stories about how humans would colonize the Moon, Mars, and beyond. But her life as the intensely shy daughter of a maid wasn't exactly *Forbidden Planet* material. Butler's mother was a widow who had no home of her own—instead, she and Butler lived in the home of her employers, a white family where Butler recalled visitors making casually racist remarks as if she and her mother weren't in the room. As an adult, Butler always expressed great admiration for her mother's tireless efforts to survive, to keep going, despite the many barriers in her way.

Perhaps for this reason, Butler's great gift as a writer was her ability to tell moving, realistic stories about how people would survive in futures far more harrowing and strange than anything that ever appeared on the *Enterprise*'s sensors in *Star Trek*. Still, she often joked that bad science fiction inspired the themes in her writing as much as growing up black in a white-dominated world. She penned her first short story after watching *Devil Girl from Mars* on TV late one night, and realizing that she could do better.

The literary world would never put Butler's work in a class with *Devil Girl*. Not only did she win many SF literary awards before her death, in 2006, including the Hugo and the Nebula, but she was the first SF author ever to win the MacArthur "genius grant" usually bestowed on fine artists and distinguished scientists. Ultimately, what makes Butler's work mesmerizing is her incredible ability to help readers see the world from a perspective radically different from their own. In an essay for *O, The Oprah Magazine,* Butler recalled a formative experience. She visited a zoo with her elementary school class, and watched in horror as the other kids threw peanuts at a caged chimp, taunting him. As the animal wailed in frustration (and possibly madness), the young Butler realized she had more sympathy for this ape at that moment than she did for her fellow humans. She'd caught her first glimpse of humanity as it might look through alien eyes, and the experience left its mark on her imagination forever. "At age 7, I learned to hate solid, physical cages—cages with real bars like the ones that made the chimp's world tiny, vulnerable and barren," she wrote. "Later I learned to hate the metaphorical cages that people try to use to avoid getting to know one another—cages of race, gender or class."

Many of Butler's novels can be understood as thought experiments in which she offers solutions to the problem posed by that group of children tormenting the chimp. At the heart of this problem are those metaphorical cages. Such cages can be more pernicious than steel bars, because they prevent humans from seeing what we have in common as members of a species in danger of going extinct. How can humans survive in the long term when we seem to be so good at building cages? What would it take to alter the course of humanity?

These are the same questions that I'm asking in this book. We can only meet the challenges of surviving whatever the natural world throws at us by working together as a species in small and large ways. Before we understand the nuts and bolts of survival strategies, however, it's important to take a short philosophical break and think about why we're doing this in the first place. Why do we want to survive? What is it that makes life worth saving? How do we hope to improve humanity over the next million years, and what would that look like?

Using a few of Butler's science-fiction novels, we can think about some possible answers to these questions.

Surviving Is Always a Compromise

Most of us want humanity to survive for a simple reason: We hope there's a chance for our families and civilizations to endure and improve over the long term. The problem is that we have a hard time imagining what that would look like. We envision a far-future world full of people who look just like us, zinging around the galaxy in ships that are basically advanced versions of rockets. And yet, if history is any guide, the humans of tomorrow will be nothing like us—their bodies will have been transformed by evolution, and their civilizations by the kinds of culture-changing events that have already marked human history. In her trilogy of novels called *Lilith's Brood,* Butler dramatizes why some people choose death over survival. They are not prepared to deal with the radical changes required to bounce back from extinction. Still, Butler's story offers us hope for humanity's survival, and a new way of thinking about how we'll do it.

When *Lilith's Brood* opens, a civilization of bizarre, tentacle-covered aliens called the Oankali have just kidnapped the tattered remnants of humanity after a nuclear apocalypse. Unlike humans, who evolved to use machine technology, the Oankali's entire civilization is based on biology. They journey through the galaxy in living spaceships the size of planets, and every part of their environment—from their tree homes to their slug-like cars—is alive. They're an ancient species who have dealt with many alien cultures, and they view humans as a fascinating anomaly: We're intelligent creatures who live hierarchically. Apparently this is an incredibly rare combination in the universe, and they suspect it's what led to our downfall. Luckily, as a representative of the Oankali explains to the protagonist, Lilith, they've preserved the few remaining humans in stasis pods while the Earth returns to a healthy state of nature.

Though seemingly benevolent, the Oankali do want something in return for rescuing the remaining humans. They awaken Lilith before all the other people to offer a bargain: They'll grant humans a rich, disease-free life if they agree to have children with the Oankali. It turns out that the Oankali evolve as a species by merging their DNA with other species, creating an entirely new kind of life every few generations. As the Oankali's reluctant ambassador, Lilith must explain the deal to her newly awakened fellows and get their consent. Some of the humans are more willing than others, but all of them are suspicious of Lilith's position—they see her as compromised because the Oankali have already reengineered her to be stronger and more intelligent than an ordinary human. Her capabilities are just a taste of what her half-Oankali children will have. But are the Oankali making the humans better, or robbing them of their humanity? Are they asking the humans to join them as equals, or to become their breeding stock?

One group of humans rebels against the Oankali, refusing to join them and opting to face death rather than form families with creatures they see as hideous oppressors. Lilith, meanwhile, consents to the deal. She and her lover, Joseph, form a typical Oankali family, which consists of a male, a female, and a third sex known as the ooloi. The ooloi can combine genetic material in its body and create mixed-species offspring which would never be possible via the kind of sexual reproduction humans are used to. Though Lilith comes to love her ooloi Nikanj, and her hybrid children, she is plagued by doubts. Maybe the separatist humans are right to refuse the bargain. Maybe the Oankali have pushed her toward accepting them by controlling her neurochemistry, slowly robbing her of the desire to resist assimilation. There's also the nagging question of whether she's truly surviving at all, if her children will no longer be properly human.

As the series goes on, these questions become even more thorny. We discover that the Oankali plan for their hybrid children to travel the universe in a living ship whose body will grow by consuming the entire Earth. Though the Oankali have, after long argument, given the separatist humans a refuge on the rejuvenated planet, this is only temporary. The unassimilated humans will die as the ship comes to life.

In some ways, the Oankali are giving humans what we've always wanted: perfect health, long lives, plenty of food, and a perfectly peaceful existence. But their bargain begins to sound a lot like what Europeans offered natives when they arrived on American shores in the wake of the great pandemics that were decimating their populations. In exchange for a few valuable commodities like guns and wool, Europeans disrupted the natives' cultures and completely transformed the lands where they lived. The longer the natives lived among Europeans, the less they seemed like Apache or Inca, and the more they seemed like hybrid peoples with one foot in their parents' cultures and one foot in their colonizers'. Even though Lilith and her children will survive, humanity as we knew it will not.

The thread that runs through *Lilith's Brood* is the idea that human survival involves radical transformation. At the same time, Butler offers us reassurance that though our bodies may change and our cultures fall under alien influence, we will retain our humanity. As Lilith's children come of age, we begin to see the world from their perspectives as creatures who are part of a species that never existed before. Though they are half Oankali, they treasure their human sides, too. Indeed, the first human-Oankali hybrid ooloi winds up falling in love with a man and a woman from a separatist human community, and discovers in the process what makes humanity so valuable. Unlike other species the Oankali have assimilated, only the humans have put up organized resistance to assimilation. As a result, the Oankali realize that they have to change their way of life. They will no longer assimilate whole species, but instead leave part of each species behind to continue on its own path. You might say that humans inject pluralism into the Oankali culture. And the Oankali, for their part, give humans a peaceful future among the stars.

So how can such an outlandish story shed light on our future as a species?

The strength of *Lilith's Brood* as a thought experiment lies in Butler's suggestion that human survival means an endless and increasingly profound series of compromises. Importantly, the books do not have a tidy, happy ending—far from it. Though the humans survive, both as pure humans and as hybrid Oankali, they endure incredible losses that some might argue are worse than death. To put this in the kind of historical perspective that we began with, the long-term outcome of cultural meetings between Africans and Europeans could hardly be described as unambiguously good, even though slavery was eventually abolished. We cannot ever hope to reach a future where the scars of history completely vanish, nor can we expect that we won't be wounded again in the future. The key is to understand those injuries in the context of a much longer story about the great transformation known as survival. Hopefully, the rewards of seeing our half-alien children building an improved world can offset the injuries that produced them. This is why we survive, Butler suggests. We want to witness the birth of something better.

In *Lilith's Brood,* Butler resists offering a pat definition of what "something better" might be. Certainly it seems that the human-Oankali way of life will be healthier, more sustainable, and more peaceful than ours is today. The author also hints that it will involve preserving what's best about humanity: our ability to change while remaining true to what came before us. Perhaps most important, "becoming better" doesn't mean transcendence. Though her future humans are vastly more powerful than us, they don't achieve a state of perfection. They are the hybrid result of compromise—better than we are, but still dealing with conflict and disappointment.

One of the great lessons about future survival that we can take away from *Lilith's Brood* is that it will require us to change. And those changes may be a lot more difficult, and a lot weirder, than we expect.

"God Is Change"

It's easy to say that we need to change to survive, but how do you get people to risk everything to do it? How do we unite people divided by those symbolic cages and work on a long-term goal together? That's a question Butler tackles head-on in two of her most realistic novels, *Parable of the Sower* and its sequel, *Parable of the Talents,* both set in a near-future United States that has been torn apart by poverty, climate change, and political instability.

We begin with Los Angeles burning down. In *Parable of the Sower,* we find ourselves in one of the last remaining gated communities outside L.A., where gangs have breached the walls and are setting houses on fire. A teenager named Lauren Olamina heeds what her father taught her on the shooting range, grabs her gun and emergency backpack, and flees into the burning night to find a safe road up to Northern California. She's heard things are better up there. Along the way, though, she and her traveling companions are kidnapped by a militia and tortured in reeducation camps. After months of beatings, they're released when the U.S. government begins to take power back from the separatists and gang leaders who have claimed the land.

During her ordeal, Lauren solidifies a plan she's had since childhood. She will create a new religion. It will be a system of beliefs that she hopes can bring people together in empathy, preventing anyone else from ever having to endure what she did. She uses the word "God" in her teachings, but not the way most Americans would. First of all, God isn't a white guy with flowing hair, floating in the clouds. God is an abstraction, described only as "change." Lauren invokes this God to aid people who are suffering, but she also claims her God is devoted to shepherding the children of Earth into space, where they will scatter joyfully to the stars. Looked at from one perspective, Butler is drawing from the Judeo-Christian God, whose idea of justice in the Bible helped African-Americans protest slavery and inequality in the United States. But looked at from another perspective, this abstract God of change reflects the idea of evolution in action. Either way, Lauren's God is a powerful idea, one that her characters in postapocalyptic America use to survive an ordeal that nearly destroys humanity.

In *Locus* magazine, Butler explained:

> I used to despise religion. I have not become religious, but I think I've become more understanding of religion.... Religion kept some of my relatives alive, because it was all they had. If they hadn't had some hope of heaven, some companionship in Jesus, they probably would have committed suicide, their lives were so hellish. But they could go to church and have that exuberance together, and that was good, the community of it. When they were in pain, when they had to go to work even though they were in terrible pain, they had God to fall back on, and I think that's what religion does for the majority of the people.

The *Parable* novels are, in essence, a story about reconciling religion with social change, God with science, and the past with the future. In these books, Butler makes explicit what is only hinted at in *Lilith's Brood:*

Humanity's story must be one of constant change because that is one way to transmute pain into hope. Lauren's goal for humanity, and, indeed, the goal of the book that you are reading, is to get us off this crowded planet and into space. There, we can continue to change and hopefully, through exploration, learn more about how to build a civilization that doesn't lock its members into various cages that prevent us from seeing our common goals.

But, as Butler told a student attending one of her lectures, "There's no single answer that will solve all our future problems. There's no magic bullet. Instead there are thousands of answers—at least. You can be one of them if you choose to be." First, however, you must be brave enough to turn away from death, embrace change, and survive.

In the next two parts of this book, we'll explore two ways humanity will need to transform in order to survive as a species, with our histories and traditions intact, but changed enough to make our future civilizations sustainable ones. We'll begin by transforming the cities where so many of us live and work. And, ultimately, we'll start building those cities beyond this dangerous, explosive planet we call Earth. We'll scatter to the stars, changing ourselves in order to survive, but always remembering home.

PART IV HOW TO BUILD A DEATH-PROOF CITY

14. THE MUTATING METROPOLIS

WE'VE SEEN HOW life-forms like cyano, birds, and mammals made it through mass extinctions, and we've explored the strategies humans used to deal with threats to our species. But we've also seen a lot of failure modes that consigned whole ecosystems and classes of people to death. How will we convert our guardedly hopeful stories of a human future into a real-life plan for survival that avoids some of the worst failure modes?

We'll start by changing our cities, which are a powerful expression of human symbolic culture and a perfect example of why we have a lot to learn about adapting mass societies to our environments. Cities have always been central to human civilization, but now they've become almost indistinguishable from it. Certainly they're the sources of our greatest economic, scientific, and artistic productivity. They're also a good way to organize communities when you're an invasive species with a population that just passed the 7 billion mark. It's easier to provide standard levels of good hospital care, sanitation, housing, and education to 1.6 million people packed into the island of Manhattan, for example, than to the less than 1 million spread out over the state of Montana. But cities are also a problem. They're death traps during pandemics and natural disasters like the 2004 Indian Ocean tsunami. Though cities are efficient in their use of energy, they still use far too much of it—especially given that most of them run exclusively on fossil fuels that are not sustainable and harm the environment.

Still, cities have become the dominant form of human community today. In the past decade, the number of people on Earth living in cities surpassed those living outside of them. And those numbers are expected to rise—the United Nations' Population Division estimates that 67 percent of humanity will live in urban areas by 2050. Certainly it would be better for people and the planet if we could dramatically decrease our population, as Alan Weisman argues in his book *The World Without Us*, but that idea simply isn't pragmatic in the next few decades. It would require us to regulate the bodies of billions of women, leading into a morally gray area from which we might never return. For now, we must accept that our population is growing. And that means human survival in the near term depends on whether we can build cities that protect their masses of inhabitants while also preserving and sustaining the environment. In short, we need cities that don't collapse at the first twitch of an earthquake, that aren't hives of disease, and that offer sustainable energy and food sources to their citizens.

To get there, we must first understand how cities work and what makes them survive over the long term.

The City Is a Process

A city is more than its brick and mortar. It is the sum of its cultural history. That's why the urban planning philosopher Jane Jacobs, in her groundbreaking 1961 book *The Death and Life of Great American Cities*, makes the case that what attracts people to cities is "sidewalk life." By that, she means the everyday social world of the city, the comings and goings of neighbors, strangers, and events. The city, Jacobs believed, was profoundly social. People flocked to them for the excitement of new kinds of human interaction, not to admire great works of monumental architecture or simply to make money.

Jacobs's interpretation is just one of many ways to express a certain ineffable aspect of city life. Some call it an emergent property, a system of organization that arises spontaneously out of chaotically interacting parts. Others call it a cultural legacy. And the fantasy author Fritz Leiber dubbed it "megapolisomancy." The point is that cities draw their vitality from a mix of social, political, and cultural practices that are hard to quantify scientifically. But they are also undeniably products of technology and engineering. They are financial powerhouses too, fueling their inhabitants' cultural and scientific undertakings in massive, elaborate marketplaces that link cities to each other across the world.

Successful cities are what physicists might call stochastic, meaning a structured, repetitive process that contains an element of randomness. Certain structures appear in cities again and again. Even the very earliest cities, located in what are today Turkey and Peru, contain monumental architecture in honor of religious and political leaders. They also contain private homes where people lived in family units, cooking, sleeping, and raising children together. And yet every city has its own character, its own random, emergent sensibility that's a product of a particular group of people at a particular time in history. Some cities, like Istanbul and Paris, manage to nourish this stochastic process over centuries and even millennia. Others, like Detroit, flash brightly for a few decades and then crumble into ghost towns. To make our cities long-lived, shaping them into "battle suits for surviving the future" as the industrial designer Matt Jones calls them, we have to respect their stochastic natures. We must build cities with safe, sustainable structures, but always leave room for randomness and social change.

Anthropologist Monica L. Smith, who researches the development of cities in the ancient world, has noted with some frustration that there is really no good way to define what makes an area "urban." Key components of urban life include a high population density, specialized forms of work, social stratification, and monument building. But listing the ingredients of a city doesn't adequately address the problem of definition, because cities are what Smith calls "a process." The brilliant urban planner Spiro Kostof suggested the same thing, writing that "a city, however perfect its initial shape, is never complete, never at rest." In other words, a city is always shifting, perhaps possessing some aspects of urban life at one time and other aspects later on. Moreover, what felt urban 5,000 years ago probably wouldn't feel urban today—and indeed, what feels urban in

Canada might not feel urban in China. We may know a city when we see it, but the idea of a city is itself a moving target.

Cities were born in two very different regions of the world: along the coast of Peru in South America, and in the area once known as Mesopotamia, where southern Turkey, Syria, and Iraq stand today. The Peruvian cities, clustered along mountain rivers speeding toward the sea, date back to 3200 BCE. They boasted large sunken plazas surrounded by platforms, winding stairs, and rooms that were probably living quarters. The largest of these cities is believed to be Caral, which dates back to 2800 BCE and may have housed up to 3,000 people who left behind art, carvings, and woven textiles. Most likely, Caral and its outlying cities lived on fishing and agriculture, trading goods and ideas back and forth.

Unlike the cities around Caral, which were elaborately planned around large, central public spaces, the even more ancient city Çatalhöyük, in southern Turkey, looks more like a honeycomb made out of mud. Anthropologists believe it was probably constructed in roughly 7500 BCE, and inhabited for hundreds of years after that. The people of Çatalhöyük built their simple one-room houses right next to each other, with no streets in between. Doors were built into the rooftops, and residents clambered across each other's roofs and down ladders to reach their pantries and beds. When the mud walls of a Çatalhöyük house began to crumble, residents would just build a new structure on top of the old one. Many ancient cities are called "mounds" because over time, ancient city people literally created hills by building new homes upon the ruins of the old. Anthropologists today often find the remains of these cities by using satellite photos to look for suspiciously symmetrical mounds in regions of the world known for early urban development.

One of the great debates among anthropologists is whether urban life or agricultural life came first. Though we may never know the answer—and it may have varied from region to region—most anthropologists today agree that cities like Caral and Çatalhöyük would have required people to develop highly efficient agriculture. After all, farming is only necessary when there are hundreds or even thousands of hungry mouths to feed in one permanent location. Did the city therefore predate the farm? Tanta-

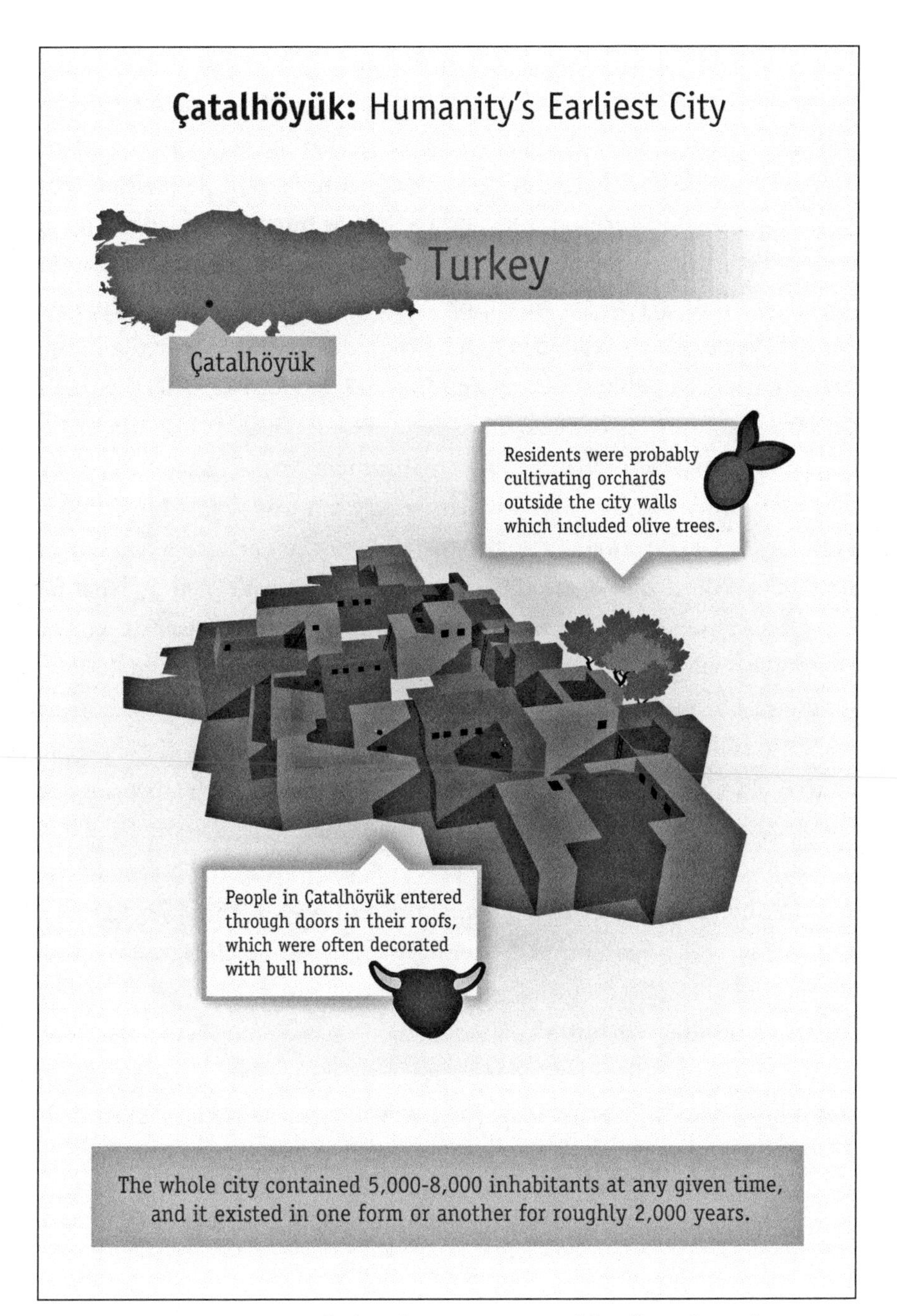

An artist's conception of what the ancient city of Çatalhöyük might have looked like. There were no streets, and people entered their homes through holes in the roofs.

lizing evidence from a southern Turkish site called Göbekli Tepe, dating from 10,000 BCE—a fascinating circular formation of monuments covered in bizarre human-animal imagery—suggests that the very earliest urban formations were built before evidence of agriculture. Still, it's possible that humans' first efforts at crop cultivation would be impossible to distinguish from wild plants, meaning the people who visited Göbekli Tepe might have had small farms that we simply can't recognize from their remains. Debates aside, what's certain is that by the time people were living in the extremely ancient cities of Caral and Çatalhöyük, farming was the main occupation of most urbanites. Cities cannot exist without agriculture.

Cities didn't just change the environment with agriculture; they changed humanity, too. Stanford University anthropologist Ian Hodder, who has led excavations at Çatalhöyük since the early 1990s, believes that cities "socialize" people. Their routines are transformed by what he calls "bodily repetition of practices and routines in the house," as well as the "construction of memories." He writes about one house in Çatalhöyük whose residents rebuilt the structure six times over a couple of centuries, each time with exactly the same layout. Like their neighbors, these people shared a religious tradition of burying the bones of their ancestors in the floor of the house. As time passed, the house became more than just a dwelling. It was a monument to previous versions of the house, to the family, and to the city itself. This is a useful way to think about cities in general, and helps illuminate why we attach so much significance to preserving ancient structures in our modern cities. Our cities are monuments to our shared history. Though the bones of our ancestors aren't literally built into the floors of our homes anymore, they remain there in a symbolic sense. That ineffable megapolisomancy that gives cities their allure comes from the way they are constructed of memories as much as they are constructed from brick and steel.

Cities That Endure

Anthropologist Elizabeth Stone has been excavating ancient cities in the Mesopotamian region, especially Turkey and Iraq, since the early 1980s.

When I asked her why some cities manage to survive for thousands of years, she cautioned me that cities don't ever remain the same over time; they have broken histories, collapsing and rising again. Ancient cities, for example, were organized in a way dramatically unlike cities of today. "If you look at pictures of Baghdad today, you see different districts that are segregated by class. It's so fundamental that it's visible from space," she said. But if you look at the layout of ancient Pompeii, it's impossible to say where the rich and the poor lived. There is variability between neighborhoods, but there are no visible differences in wealth. As she's mapped Mesopotamian Era cities, Stone has been struck by how little variability there is in the size of houses. Everybody seems to have homes that are roughly the same dimensions, though some might have more rooms than others.

The differences between ancient and medieval cities are just as stark. The imperialist Rome of antiquity wasn't the same as the Church-dominated Rome of the Middle Ages. The former glory of the ancient world was reborn as a new city for the medieval world. Medieval city growth moved slowly, often funded by the aristocracy or the church. But starting in the nineteenth century, industrialization pushed city growth into the hands of wealthy entrepreneurs and developers, whose greatest monuments became skyscrapers devoted to various corporate headquarters. This era also witnessed a steep rise in urban populations, culminating in our majority urban population today. And now it's become a completely different city again. People have always been drawn to Rome because of its dramatic history, but the urban experience during each stage of its life was notably transformed.

Long-lived cities survive by going through periods of collapse and rejuvenation. It's possible that cities tend to collapse when people have less social and economic mobility. "People at the bottom may retreat into the countryside and leave the sphere that's controlled by the city," Stone speculated. As soon as there is more opportunity in the city, peasants return and try to climb the social ladder again. Most cities that last for more than a few hundred years are located at the heart of shifting empires, like Istanbul (formerly Constantinople) or Mexico City (formerly Tenochtitlán). Both cities have been inhabited for centuries, but by peoples from successive, often adversarial political groups. Their fates rise and fall with the empires that claim them. Cities may be built on memory, but they are also processes, always changing.

Longevity isn't the only measure of a city's success, however. As Harvard economist Edward Glaeser puts it in his book *Triumph of the City: How Our Greatest Invention Makes Us Richer, Smarter, Greener, Healthier, and Happier:*

> Among cities, failures seem similar while successes feel unique.... Successful cities always have a wealth of human energy that expresses itself in different ways and defines its own idiosyncratic space.

Modern cities survive by offering people a space where they can form social groups that would be impossible outside them. Kostof calls cities "cumulative, generational artifacts that harbor our values as a community and provide us with the setting where we can learn to live together." The city community's "values" are part of the urban structure itself. This helps explain why some cities remain politically distinct from their countries, their cultures proving stronger than the culture of their nations. In the last century, West Berlin and Hong Kong found themselves in this situation. Both cities had strong ties to other nations and urban areas, and used those ties to remain relatively democratic cities devoted to capitalist trade, despite being located inside and alongside powerful communist nations. Other examples include cities like New York and Budapest, whose citizens have often defined themselves by being at odds with the countries that contain them. Cities socialize citizens into certain habits of mind, and these can be hard to break. In fact, sometimes a city breaks with its nation rather than breaking away from its own social norms.

Case Study: San Francisco as a City of Tomorrow

How can we ensure that tomorrow's cities will harbor thriving communities that don't decay into insignificance like Detroit or disappear into the fog of history like Çatalhöyük? We need to incorporate mutability into urban design. But as we face the future, that mutability must also include ways of building sustainability into the very structure of our cities. Urban geographer Richard Walker believes the San Francisco Bay Area provides a useful template for how that could happen. In his book about San Francisco, *The Country in the City*, he explains how the Bay Area's green spaces are as much constructions as the houses and buildings. The region's designers often built verdant parks on top of barren dunes and scrub, including both "country" and "city" in their plans for how they would convert the wild lands of Northern California into an urban space. The results are visible everywhere in the Bay Area. To get from the BART commuter-train station to Walker's Berkeley home, I followed a winding path through several public parks full of play equipment and flower beds. Along the way, I passed more bicyclists, pedestrians, and green spaces than I did cars.

Still, the Bay Area isn't built on environmental principles alone. It's also a successful region because its residents have consistently been on the cutting edge economically. "Going back to the Gold Rush, San Francisco has always had a big skilled labor force full of young, creative people," Walker said. "That was true when they were inventing new kinds of mining equipment in the nineteenth century, and it's true now with financial innovation and retail innovation, as well as electronics and biotech." The Bay Area's financial heart is truly in the long, braided terrain of farms, parks, and cities that make up Silicon Valley to the south. They pump cash from the innovative tech and science industries into a region that includes Marin County to the north and Berkeley and Oakland to the east. Today, many young people who are attracted to the culture of San Francisco come to the region to settle in its most famous city. But they commute every day to Silicon Valley in one of hundreds of sleek, Wi-Fi-enabled buses dispatched by Google, Genentech, Apple, and other companies to make commuting easier on their employees—and reduce emissions in the process.

Just as important as its economic success, however, is San Francisco's status as what Walker called a "wide-open city." By that, he means a city prepared to tolerate, and even embrace, experimental ideas. In the early twentieth century, the Bay Area was home to the nation's earliest environmental groups, racially integrated unions, and a large gay community. In this way, San Francisco in the 1920s and '30s was like Los Angeles and Berlin. But unlike those cities, the Bay Area never suffered a political crackdown on its most rebellious citizens. During the rise of fascism in Berlin, the Nazis drove out (and occasionally murdered) progressives and openly gay activists like the psychologist Magnus Hirschfeld. And in 1950s Los Angeles, the House Un-American Activities Committee persecuted people with leftist sympathies working in Hollywood. Many lost their jobs and had to leave the city. Meanwhile, in San Francisco, the radicalism continued virtually unchecked. Citizens founded environmentalist groups like the Sierra Club, and a general strike in the 1930s brought the city's bosses to their knees. In the 1960s, an odd set of local environmental groups, industrialists, and politicians came together to battle developers who wanted to top up the bay with landfill so the city could sprawl from the Embarcadero in San Francisco across to Alameda in the east, and all the way down to Redwood City in the south. The environmentalists won that round, and the bay was protected from destruction.

Out of struggles like these arose a city unlike its predecessors, an urban environment that was green almost from its very inception. "The environment in the Bay Area welded many local opposition movements into a larger radical vision of a green city," Walker explained. "The feeling here wasn't 'protect my neighborhood and screw everybody else.' It was 'protect my neighborhood and come hike in my green space.' It was very public-spirited." By the mid-1960s, the city's coalition of local green groups had become so powerful that California passed the first of many environmental-protection laws to prevent anyone from ever filling in the bay for development.

In building the Bay Area, urbanites realized that success meant destroying the false division between country and city. But San Francisco is just one example of a city that has changed over time by getting greener. People in many cities, from Tokyo to Copenhagen, want to preserve local environments not just with protection laws, but also with solar power, high-efficiency buildings, and urban farms. Cities of the future are changing to include many aspects of the country within their boundaries.

As we'll see in the next few chapters, urban planners, architects, and engineers are coming around to the idea that cities must be as much part of their environments as coastlines and trees are. Government and private industry are pumping billions of dollars into the development of energy-efficient buildings, solar power, smart grids, urban gardens, green roofs, and many other eco-technologies. The city of the future, most agree, will be planned the way the Bay Area has been for almost 50 years.

It may seem bizarre for the Bay Area to represent urban life of the future, given that an enormous earthquake or tsunami could wipe out the whole region tomorrow. But as we'll discover in the next chapter, new engineering techniques could help our cities survive all but the worst natural disasters.

15. DISASTER SCIENCE

THERE'S ONE THING that never changes when it comes to city life. Disaster will always strike. Whether it's from storms, floods, earthquakes, fires, or just urban decay that's turned buildings into deadly hulks of rotting wood, cities fall apart. One of the biggest questions for urban planners and engineers is how to build cities that can withstand common calamities. It turns out the best answer is to destroy a lot of buildings on purpose. Engineers innovate city-building technologies by using enormous labs to re-create the worst disasters you can imagine—and then inventing structures that survive them.

Many of these labs are in remote facilities that you might at first mistake for storage warehouses, missile ranges, or airplane hangars. Several years ago, I crisscrossed the United States, trying to visit as many disaster labs as I could. I started with the Energetic Materials Research and Testing Center, a 40-square-mile swath of blue-veined rocky hills covered in sage brush next door to the White Sands Missile Range in Socorro, New Mexico. Between peaceful hillsides mostly dominated by wildlife, researchers from New Mexico Tech collaborate with government and industry scientists to study how explosions affect city environments. The day I was there, emergency responders set off a car bomb to see whether a specially reinforced brick wall could protect a test dummy from the blast.

The dummy survived, though the "control" dummy behind a standard wall was shredded, as was the car. Analysts pored over the crater the car left behind, measuring the distance that the engine had traveled, trying to analyze every factor in the explosion. Other tests at the facility measure the effects of tanker explosions, gunfire, and even tiny suitcase bombs. Their results could help city planners and rescue workers design streets and walls to protect residents from harm.

Tests like these also help rescue workers learn new ways to pull people from wreckage that can be even more dangerous than the blasts that created it. Rescue innovation is a big part of what scientists and emergency responders study at Texas A&M's Disaster City, another enormous open-air facility devoted to destruction for the sake of survival. Here, engineers can build whole city blocks just to blow them up in a re-creation of a meth-lab explosion or a house fire. They can simulate a train crash or root around for survivors in a collapsed parking structure. When I visited, engineers were testing experimental reconnaissance robots designed to fly or climb around in dangerous, unstable environments to find people trapped in rubble. Next to Disaster City is a fire field with a mock chemical-processing plant. While I watched, the technicians opened the valves on gas lines that fed into a maze of pipes and tanks, emulating what would happen if such a plant caught fire. Firefighters struggled to contain the two-story flames. I stood in the heat-mangled air outside the painted safety lines that bracketed the area like the sidelines on a basketball court.

While these facilities specialize in pyrotechnics, another network of labs in America and Japan are filled with huge machines that can simulate earthquakes and tsunamis. At Oregon State's tsunami lab, engineers carefully erect scale-model cities around the "shoreline" in a 160-by-87-foot water tank, then create carefully designed tidal waves with huge paddles to see where the water washes ashore. The tank is lined with sensors that measure the movements of tiny beads of glass suspended in the water—this allows researchers to understand how waves propagate through oceans, and better predict how tsunamis will behave when they hit the shore. Sitting high above the tank in a control room, scientists use a computer to control the paddles, generating exactly the kinds of waves they want to send crashing down on the model city. They can imitate the conditions that would affect the speed and shape of a tsunami in a very specific region, such as the northern coast of Oregon or the San Francisco Bay. Ultimately, these tests help city planners determine a safe distance to build from the water, as well as the optimal places to put escape routes in case of flooding.

Researchers at the Oregon State University Tsunami Lab have built a model city in the football-field-sized wave tank and use enormous machine-controlled paddles to generate a wave. Simulating a tsunami disaster will help them plan better how to build cities that can withstand flooding.

As scientists in these labs struggle with floods and fires and quakes, they are also struggling with a fundamental contradiction at the heart of city design. As the urban planning historian Spiro Kostof explains, cities are the result of ongoing conflicts between centralized planning and organic, grassroots development. To prevent people from dying in quakes and floods, for example, we need rules about how and where developers are allowed to build. But city governments can't control everything. City dwellers aren't going to be happy if they don't have the freedom to change their living spaces and neighborhoods. Not everyone can afford to build homes that are robust against every kind of possible disaster, either. That's why engineering a disaster-proof city isn't about magically conjuring damage-proof structures. Instead, it means building urban areas that will kill the smallest number of people possible during a disaster. This is pragmatic optimism at its most literal.

Inside the Apocalypse Lab

I met the UC Berkeley civil engineer Shakhzod Takhirov inside a three-story warehouse that's home to UC Berkeley's Earthquake Simulator Lab. Located in the city of Richmond, the lab is easily identified by its proximity to piles of shattered wood beams, twisted girders, and giant cracked columns of concrete. But this was no junkyard. As I wandered through the rubble, I noticed that every crack and break had been carefully labeled with measurements in permanent marker.

The instruments of destruction that created these piles occupy most of the lab. Towering over my head as I walked in was a 65-foot-tall steel piston that can deliver up to 4 million pounds of compression to whatever structure or material is unlucky enough to be in its grip. Want to simulate traffic load on a bridge support, or the pressures that a skyscraper might deliver to its foundation? This machine can help.

Behind the mega piston, I could see that day's main experiment. Lab technicians had built a life-sized frame for a single-story building in the middle of the warehouse-sized lab space. Attached to the frame were huge hydraulic motors that looked a bit like pared-down robot arms that were braced between the building and a strong concrete wall. These motors were controlled by researchers in a room packed with computers. With the press of a button, the engineers could deliver small, precise earthquakes to the building—or bone-rattlingly big ones. Sensors on the structure would measure every deformation and shake propagated through it.

Takhirov, who bounced around the control room taking obvious delight in the powerful machines working outside, has always lived with the threat of earthquakes. His birthplace in Uzbekistan is known for its

In this image from the Earthquake Simulation Lab at the University of California, Berkeley, researchers were experimenting with the performance of steel-frame buildings reinforced with a variety of braces. They tested several braced frames and simulated earthquakes with large hydraulic actuators (the black objects with cylindrical bodies on the right side). Each actuator was capable of applying up to 1.5 million pounds of force.

massive quakes, as is the San Francisco Bay Area, where he's spent much of his adult life. Though he began his career as a mechanical engineer studying wave dynamics, over time he left theory behind and got interested in real-world applications. The day I visited, the researchers were deforming one wall of their building with the motors. The process was slow, involving tiny shifts in the slightly crushed structure, and a great deal of muttering from graduate students about the waveforms we could see undulating across several computer monitors.

There are two ways to simulate earthquakes. Researchers can use a shaking table, which is exactly what it sounds like. They build a structure on top of a platform that can be shaken from underneath, creating an earthquake, so they can watch what happens and learn from it. The second way is what Takhirov's colleagues and students were doing. They used their giant actuators to imitate how earthquake forces would deform the building, but they were doing it in slow motion. There was none of the violent motion you would see in an earthquake, but those robot arms carried the same force as a quake would. "Essentially we do this so we can look at each step," Takhirov said. Using their computers, the researchers can also create a "hybrid simulation" that combines a mathematical model of a building with the physical object they're manipulating in the lab.

The experiment that I was watching with the one-story building turned out to be a model of a two-story building—the second story existed only in software. We know enough about earthquake engineering at this point that we can actually extrapolate how a second story might behave based on what the first story does when it is slowly crushed by giant motors. Hybrid simulations make it easier for engineers to calculate how city buildings might respond in a quake, even if they aren't able to build an entire 50-story building and wiggle it.

This particular hybrid simulation would ultimately reveal what happens to a multistory building during a quake if the second floor had been "isolated," or built with a damper—usually a layer of flexible material—between it and the first story. Isolation stops the quake's motion from propagating through a building unchecked, preventing it from swaying, torquing, and crumbling. Usually isolators are built into the bases of buildings, but the experiments I saw would demonstrate whether isolation units could be helpful between stories, too. If the isolator prevented significant damage in that simulated second story, these researchers would move on to the next phase of their work–getting their engineering discovery implemented in the real world.

How to Prevent Death with Engineering

What Takhirov and his colleagues learn in the Earthquake Simulator Lab gets translated into the building code, a set of safety rules that constrain how structures are built. These codes exist all over the world, varying slightly from region to region. When engineers like Takhirov make a new discovery about earthquake engineering, their next step is to petition to change the rules that govern city development. "I can conduct several tests, and then I can approach the coding committee with my results and say, 'I should change things here,' " Takhirov said.

Failure to update building codes is a major reason so many lives are lost in cities during disasters. Takhirov visited Haiti soon after the series of quakes that nearly leveled the capital, Port-au-Prince. He and his team documented the damage, using conventional cameras as well as sophisticated laser-imaging devices that produced 3-D representations of the shattered city. A lot of the damage could have been prevented with better engineering. They found buildings that never would have collapsed if they'd used simple reinforcements. Unfortunately, however, the local building code lagged behind recent discoveries. But some buildings weren't up to the local code, either–mostly because builders couldn't afford the reinforcements and structural planning required. The more earthquake-proof a building is, the more expensive it gets. That's why Takhirov tries to be pragmatic about earthquake engineering. When builders have to cut corners, they should always prioritize human safety over a building's durability. "Sometimes it's more cost effective to have a building that will be damaged but not collapse," Takhirov explained. "That way people can escape, even if you have major damage."

His thoughts turned to what Bay Area residents call the Big One, or the next massive earthquake that could hit the region at pretty much any time. "We must be aware that the Big One will be strong, but I have some confidence that it's going to be okay, and a minimal number of lives will be lost from collapsed buildings," Takhirov said. Still, he wasn't sanguine. "Unfortunately, the Big One is going to happen no matter what," he said. And then, like a true engineer, he began imagining the discoveries such an event would yield. "When it happens, we will deploy all our cameras, and that will be our next big project."

Other engineers are more fatalistic than Takhirov about how many lives they can save. One state north of Takhirov's earthquake-simulation lab, on a hillside in the middle of Oregon's Willamette National Forest, a U.S. Geological Survey (USGS) engineer named Richard Iverson has created hundreds of landslides to learn more about how these often deadly disasters start. He does his work at the USGS "debris-flow flume," which is pretty much what it sounds like. It's an outdoor laboratory that consists of a massive enclosed slide, adorned with cameras and embedded with sensors that measure everything from pressure to sheer force while fast-moving globs of mud, rocks, and water rush down the slope. When I spoke to him, he'd just finished a series of experiments where he and his colleagues sent debris flying into mud dams at the base of the flume. They were imitating a common and deadly scenario, where a mudslide temporarily dams a canyon, water builds up behind it, and then homes below are destroyed when the whole mess breaks open in a terrifying flood. After each experiment, Iverson feeds the data he's gathered into predictive models, or computer programs that forecast disasters based on current conditions. Already, he said, he and his colleagues had learned more about flood warning signs after mudslides.

Research at the flume has led to an extremely sophisticated warning system on Mount Rainier in Washington. Several communities on the mountain suffer from periodic landslides due to water runoff, but Iverson and his colleagues were able to plot where these slides were most likely to start. They set up a warning system, a network of sensors that get tripped when a landslide's characteristic ground vibration begins. When that happens, an alarm system is immediately set off and residents below get 30 to 45 minutes' warning so that they can escape an event that often ends in death. Engineers at the flume also tested specialized wire nets that now lie like spiderwebs across the hillsides and cliffs above many highways in

California, preventing small landslides from spilling onto cars or blocking the road.

Still, Iverson said, he feels like people in the United States don't think enough about natural disasters when building cities and towns. "Some places really do make use of our predictions for guiding future development," he replied. "But frankly, in the United States, with our history of zoning laws and development, it doesn't get taken into great account." He said that the big problem is that a lot of risky areas, such as the flood-prone Los Angeles canyons, were built up before anyone knew about the danger of mud slides. "You don't always get to change much, so you do what you can."

Ideally, Iverson said, he and his team would have enough resources to get detailed topographical maps of any part of Earth so that they could run landslide models of them and determine the safest places for people to build. "We could create probabilistic models for any area that we had data for, showing a range of possible events, from very likely and not so bad, to unlikely and very bad. Showing this information on maps would be very useful for planning purposes." With the right amount of data, Iverson believes, he could give any planner a fairly realistic prediction about whether future cities might be in danger of getting buried in mud slides the next time a storm hits.

By destroying buildings and causing mud slides, Takhirov and Iverson are able to study disaster as scientifically as possible. What they've learned has already affected how cities are built, and how people evacuate flood zones. As we move into the future, however, we want cities that can do more than collapse without killing us. We want cities (and city emergency services) that can change instantly in response to imminent danger. Such cities, though they sound like science fiction, are already in the process of being designed.

Smart Cities and Disaster Prediction

City planners looking to the future often talk wistfully about data acquisition. With enough data about how natural disasters have unfolded in the past, prediction becomes much easier—especially when computers are involved, juggling thousands of data points every nanosecond to create a likely model of the future. That's why IBM recently launched its Smarter Cities program, which is essentially a suite of software and services that the company sells to cities whose governments want to predict everything from traffic and crime patterns to the best exit strategy in a flood. The goal is to create cities whose traffic, food systems, energy grids, water management, and even health care are managed in a "smart" way, based on real-time data that reveals what's needed where. This "big data" can come from almost any networked gadgets, including sensors, mobile phones, and GPS devices.

George Thomas, a former structural engineer, heads up the company's Smarter Cities sales efforts, and has helped implement the program in several urban areas around the world. One of their first projects was to reduce traffic in Stockholm. First, IBM installed cameras over heavily trafficked roads near downtown to gather data. Once they had enough information, they were able to predict peak traffic hours every day. To reduce traffic, the city installed a ring of sensors around the city center that identify the license plates of every car passing through. If cars pass through during a period of high traffic congestion, drivers will automatically be charged a "congestion tax" at the end of the month. Almost immediately, the city found that more people took public transportation, carbon emissions went down, and city revenues went up. Most important, the traffic snarls around the Swedish city were gone.

One of the group's current projects uses data that engineers like Iverson have been gathering about how mud slides and floods behave. Their goal is to give residents of Rio de Janeiro two days' warning before the notoriously flood-prone region is inundated with water and mud gushing down from the mountains that ring the city. In the past, emergency responders have only had a six-hour window in which to evacuate, but that's not long enough. With the city set to host the Olympics and soccer's World Cup, Rio's mayor decided to work with IBM to create a system that could predict floods as far in advance as possible. They needed what Thomas called a city operating system—a piece of software that could integrate streaming data from sensors on local flood plains and weather monitors. Working with all this data, the city's operating system could convert many types of information into a predictive model that would change in real time. With the new system in place, people in Rio will have a full 48 hours to leave their homes and get out of the city before disaster strikes.

Of course, even the best-prepared regions are still going to suffer setbacks. Japan was unprepared for the calamity of the March 2011 earthquake and tsunami. Though the damaged Fukushima Daiichi Nuclear Power Station did have flood-protection walls, they weren't high enough. And switches that would have brought backup power to the plant's cooling units hadn't been adequately flood-proofed either. Could a predictive system with enough data have helped disaster workers prepare for the event? Possibly, though in the wake of the disaster, officials discovered that workers had known about problems at Fukushima for years without addressing them. Predictions are only helpful if city builders are willing to act on them.

The question, as always, is how to build based on what we know. Takhirov thinks the solution lies in the building code, which changes as new discoveries are made. But as Iverson explained, it's not always easy to change cities that have already been built. All we can do in those situations is to make our cities "smarter." That's why engineers around the world are gathering all the data they can about disasters that might hit, in order to offer accurate predictions about when they'll happen, and how to escape. Cities are more than buildings; they are the people who inhabit them. The philosophy of disaster science is that it doesn't matter if our structures are damaged as long as people survive. Those people will come back to rebuild the city.

As we'll see in the next chapter, engineers weren't the first ones to come up with the idea of modeling urban disasters to save lives. It's a strategy that works with pandemics, too.

16. USING MATH TO STOP A PANDEMIC

ENGINEERS WHO WANT to prevent natural disasters from destroying cities are still in the process of gathering enough data to predict dangers before they happen. But epidemiologists have been using predictive models for a century and a half. In the 1850s, a doctor named John Snow carefully mapped every incidence of cholera he could find in London, eventually determining that a single well was ground zero for the disease outbreak. It was the first great triumph of epidemic modeling, or using maps and data to figure out how infectious disease spreads through a city. Today, we have decades of data like Snow's to help epidemiologists predict future paths of infection–hopefully stopping pandemics before they spread.

The next pandemic could start with viruses. Or bacteria, the way the Black Death did. It could even start with a bird or pig; with just the right combination of genetic material, a pathogen can jump from an animal into a human host. If the pathogen is infectious enough, the pandemic could kill 50 million people, the way the Spanish flu did in 1918. If it's highly virulent, or develops resistance to treatment halfway through an outbreak, it could kill billions.

One of the most realistic pandemic scenarios in recent years came out of Hollywood, in a movie called *Contagion*. It offered a step-by-step look at how nations and health departments would respond to a viral outbreak, and how it would spread quickly via travelers all over the world. And–spoiler alert–it also gives us a very plausible scenario for the disease origin. A development company in China has been cutting down forests, displacing the local bat population. With no more natural habitat, the bats wind up nesting in barns full of pigs, and their virus-laced guano falls into the pig mush. Eventually a visiting American is exposed to one of the pigs, and she takes the virus with her back to the United States. Thus a pandemic virus is born, partly the result of human meddling in the environment, and partly the result of our cosmopolitan living arrangements.

So what are we going to do about it?

When planning to survive a pandemic, there are two basic questions: What is killing people, and how should we organize an international response to it that minimizes death and economic damage? Generally we can answer the first question in a laboratory–and often we can come up with a treatment and vaccine there, too. As difficult as it is to identify a killer microbe and come up with a way to fight it, it's the second question that keeps scientists and policy-makers up at night.

Even if we manage to whip up a cure for the pandemic, it won't do any good if we can't get it to people in time. That's why modeling possible pandemic-outbreak scenarios has become its own scientific subfield, combining everything from medicine and genetics to statistical analysis and game theory. Pandemic modelers are usually experts in mathematics, creating gamelike computer simulations to aid in predictions, as well as maps and charts representing how a pandemic might spread. They also draft charts of survival. The pandemic modeler's goal is to figure out what groups like the World Health Organization (WHO) and local medical groups can do to intervene and change the odds. They answer questions about how much vaccine we would need to prevent a pandemic from infecting a small town and how much for a large city. Using known patterns of infection, they can figure out roughly how much a quarantine will slow the spread of a pandemic, and the minimum number of antiviral drugs a country should stockpile to prevent mass death.

We already have many of the medicines we need to kill pandemic diseases. But to stop the pandemic itself, we need math. We have to understand how a pandemic is likely to unfold across the globe, in many societies, before we can set up the best system for stopping it.

The medical surveillance state

Over 10 years ago, the U.S. government asked the CIA to work on pandemic prevention. Using the country's most notorious spy agency to deal with a health-care issue sounds like a bizarre fit. But it is the perfect organization, because pandemics are prevented in part by using techniques borrowed from spying. Does that mean our survival is dependent on everybody enduring forced health checks every week during flu season? No. And it doesn't mean the government will be snooping through your medical history either. Even if the CIA wanted to dig through everybody's medical records, it would be impossible, because many people don't have health insurance or receive regular checkups. The CIA helps medical organizations craft strategies for health surveillance, or the practice of gathering information about who is coming down with infectious diseases and where they are.

The World Health Organization (WHO) and other health-monitoring groups rely on a combination of sources for their health-surveillance data, including news stories about flu outbreaks and virus samples from all over the world sent to the WHO's Global Influenza Surveillance and Response System. WHO scientists working with Google have also created Google Flu Trends, a system that monitors flu outbreaks by tracking the search terms that people are using in various regions. Google researchers discovered that when there was a significant uptick in people searching for words related to flu symptoms, like "sniffles" or "fever," it was almost always followed by the Centers for Disease Control and Prevention (CDC) identifying a flu outbreak. Now the CDC and other agencies use Google's data to figure out where the flu is breaking out, days before people start going to the doctor to report the symptoms they researched online. Like all forms of health surveillance, Google's flu data is made as anonymous as possible. All we really need to know is how many people have flu symptoms in a specific region—we don't need to know their names or their street addresses.

Though the CDC and the WHO are the organizations we think of first when it comes to containing a pandemic, the greatest asset in any surveillance network is always your local health department, where the signs of an outbreak are going to be registered first. David Blythe manages health surveillance for the Maryland public-health department, which coordinates with dozens of regional health departments in the state to track what are called flu-like symptoms. Blythe said that one of the main ways the CDC tracks potential outbreaks is with a volunteer network called ILINet (for Influenza-like Illness Surveillance Network), a volunteer effort by local doctors, nurses, and other health-care workers who report any infectious, flu-like symptoms they see cropping up in patients. It's key that they report symptoms rather than trying to diagnose what they're seeing, since one of the main things ILINet is designed to catch is a new, deadly flu strain. If one arises, its collection of symptoms may not match any known illness. Every week, analysts with ILINet pore over the data, looking for suspicious patterns. What's crucial here is that this health surveillance is happening on a city-by-city basis. Pandemics always start in one place, as John Snow found with the cholera-infected well in London. In other words, when the next big pandemic starts brewing, city health-care workers are going to notice it long before national and international agencies do.

To supplement the work of ILINet, Maryland also has a group of volunteer labs that send samples of flu strains they've collected to the state health department for testing on a regular basis. "This is a lab that's just designed for surveillance," Blythe said. "We can do the testing that tells us whether it's AH3 or N1, and we can determine if it's a pandemic strain." And if they discover a new pandemic strain, they ship it to the CDC in Atlanta. Maryland is also working on a way for low-income and homeless people to report when they have the flu as well, since they tend to fall outside the health department's surveillance network. "We know that many people with flu never seek out a health-care provider at all," Blythe lamented. The Maryland Department of Health and Hygiene tries to remedy this by asking people to report in when they or somebody they know has the flu, even if they don't go to the doctor. The agency also tracks illnesses in health-care workers, since they are often on the front lines when pandemic strikes. "If a new strain of SARS"—severe acute respiratory syndrome—"starts in Maryland and nobody recognizes it as SARS, the first place you'd see people getting sick would be hospitals, so we have surveillance to try to pick up that phenomenon," Blythe explained.

Patterns of infection

Though a deadly pandemic could arise from the flu, it might also be a product of that ancient scourge the plague. A mutation in *Y. pestis,* the plague-causing bacteria, could leave us vulnerable to one of the deadliest diseases humanity has ever confronted. We might also find ourselves battling SARS, or (less likely) a virus like Ebola, which causes the extremely deadly and infectious viral hemorrhagic fever.

Regardless of the microbe that threatens us, a pandemic proceeds through eight recognizable stages, from incubation in animals at stage 1 to full pandemic in multiple countries at stage 6 (the peak of the pandemic). The next two stages, post-peak and post-pandemic, occur when the disease ebbs away until no one is infected anymore.

Nils Stenseth, a biologist with the University of Oslo's Centre for Ecological and Evolutionary Synthesis, is an expert on plague. He and his colleagues lay out the typical scenario that most people expect for a pandemic, based on what they know of historic Black Death outbreaks:

> In this classic urban-plague scenario, infected rats (transported, for example, by ships) arrive in a new city and transmit the infection to local house rats and their fleas, which then serve as sources of human infection. Occasionally, humans develop a pneumonic form of plague that is then directly transmitted from human to human through respiratory droplets.

Like the flu scenario in *Contagion,* this pandemic starts by infecting animals and quickly spreads to humans living in cities. Though Stenseth cautions that modern pandemics don't always bloom first in cities, most pandemic modelers take cities as the fundamental points of contagion—the dots on a map that spawn red vectors of infection.

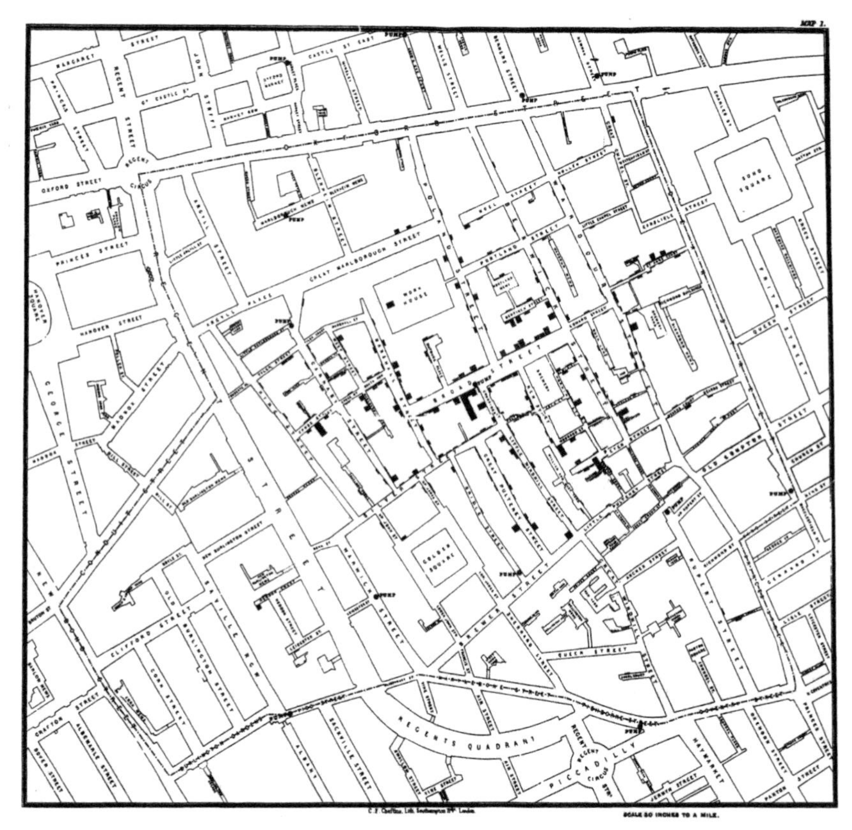

John Snow's famous map of London, in which he tracked a cholera epidemic to its source in a well. You can see where he drew lines for numbers of dead in each location. He zeroed in on the well by tracking where the greatest numbers of deaths had occurred.

But how do we predict where those red vectors will go once they've left the city behind? There's one major difference between the Black Death hitting London in the 1340s and SARS hitting Hong Kong in 2003: air travel.

Though the SARS outbreak began in mainland China, investigators with the WHO and the CDC tracked its global spread to one isolated incident at Hong Kong's Metropole Hotel. A medical professor visiting from southern China, where SARS had been claiming lives for a few months, checked into a small room on the ninth floor. Within days, 16 guests and visitors to that floor had also come down with the illness–many of them becoming sick after they'd flown to other cities all over the world, from North America to Vietnam. Investigators later came to call this incident a superspreading event, and traced it back to a hot zone on the carpet in front of that infected medical professor's hotel-room door.

Even three months after the professor had checked out of the hotel, technicians were able to find SARS viruses in the carpet. In its report, the WHO speculated that the sick professor might have vomited outside the door to his room, leaving behind a massive number of live viruses that survived a cleanup from hotel staff. Somehow, those viruses wound up in the lungs of 16 other people who passed near the hotel hot zone, and carried it all over the world–starting what nearly became a pandemic.

Incidents like the one in the Metropole Hotel have led pandemic modelers to build air-travel routes into nearly all their outbreak scenarios. Tini Garske is a mathematician and researcher with the Imperial College London's Centre for Outbreak Analysis and Modelling, and she's spent most of her career modeling disease outbreaks. Her most recent work focuses on generating outbreak scenarios based on Chinese travel patterns. She and her colleagues surveyed a group of 10,000 Chinese people from two provinces, looking at typical travel patterns in both rural and urban regions. What they found was that pandemics emerging in rural areas are likely to spread "sufficiently slowly for containment to be feasible," because most people surveyed rarely traveled outside their local areas. Economically developed urban areas make containment more difficult, owing to the numbers of people traveling great distances on a regular basis.

It would seem that the answer is simply to prevent people from traveling during a pandemic. But by the time we know we're in the midst of a pandemic, it's too late. Many other models show that limiting air travel makes almost no difference when it comes to limiting the spread of disease–at most, this tactic could delay the spread by a week or two. There are, however, a few superior methods based on models that take Garske's travel research into account, and that incorporate what we learned during the SARS near-pandemic and the H1N1 (swine flu) pandemic of 2009.

Social distancing

Usually the first strategy that comes to mind for stopping pandemics is quarantine. In a typical quarantine, the government separates people who have been exposed to the disease from the general population. Ideally, people who have the pandemic disease are isolated both from the general population and from the quarantined.

During the SARS outbreak in Toronto, the Canadian government quarantined hundreds of people, and a number of large public events in the city were canceled, in an effort to contain the disease. After the dust settled, however, many medical experts, including representatives of the CDC, argued that the local government had overreacted, quarantining roughly 100 people for every SARS case. The chief of staff at York Central Hospital in Toronto, Richard Schabas, criticized the city sharply in a letter to a Canadian journal devoted to infectious disease: "SARS quarantine in Toronto was both inefficient and ineffective. It was massive in scale," he wrote. "An analysis of the efficiency of quarantine in the Beijing outbreak conducted by the American Centers for Disease Control and Prevention concluded that quarantine could have been reduced by two-thirds (four people per SARS case), without compromising effectiveness." In other words, the mass quarantines we see in virus horror movies like *I Am Legend* are not the way to stop a pandemic. They burn through health-care resources and are ineffective.

If we're facing a brewing pandemic, however, there are good reasons to avoid large-scale social events where the disease could spread. Canceling a large concert, or asking people to stay at home, are both part of a pandemic-containment technique called social distancing. Most experts believe that social distancing and quarantine on a limited basis can help: At UCLA's David Geffen School of Medicine, biomedical model expert Brian Coburn and his colleagues claim that school closures and discouraging big public events can reduce the spread of flu by 13 percent to

17 percent. Voluntary quarantine in the home seems to work better than closing schools, though closing schools is often a sound policy because a microbe's fastest route to pandemic status is to infect children.

Vaccination must be global

As we've seen already, quarantine works in only limited doses. What's our next option? Let's consider vaccination, which many of us are familiar with from the H1N1 (swine flu) pandemic of 2009. Vaccines program the immune system to recognize and neutralize disease-causing microbes that enter our bodies. When we get flu vaccinations, we receive a small dose of damaged and dead flu viruses that help our bodies build up antibodies tailor-made to stop the flu when it shows up. Vaccines are usually not cures, and don't generally help people who are already sick; they are used as a preventative measure.

Most pandemic modelers agree on one thing: Vaccines stop pandemics only if they are administered very early in the outbreak, before the disease has had a chance to spread. Laura Matrajt, a mathematician at the University of Washington in Seattle, has modeled several strategies for containing pandemics with vaccines. The problem, she points out, is that pandemics spread differently depending on the population—a rural outbreak is very different from an urban one. They also spread differently in the developed world than they do in developing countries, largely because children make up nearly 50 percent of the population in many developing countries (in most developed nations children are less than 20 percent of the population).

Vaccinating children is vital in stopping a pandemic, because they are what Matrajt calls a high-transmission group. In other words, children are humanity's biggest spreaders of disease. If we can vaccinate kids against a pandemic disease, it will spread slowly enough that we can contain it and protect adults, too. Coburn reports that some of his colleagues found that "vaccinating 80% of children (less than 19 years old) would be almost as effective as vaccinating 80% of the entire population."

The problem is, most children are in developing countries that cannot afford to buy vaccines. This is where science butts heads with social reality. Pandemic modelers have to add dark economic truths into their equations, and figure out how best to administer vaccines in a situation where perhaps only 2 percent of the population will have access to it. Matrajt and her colleagues came up with several scenarios in the developing and developed worlds, where people had access to different amounts of vaccine, ranging from 2 percent coverage to 30 percent. "For a less developed country, where the high-transmission group accounts for the majority of the population, one needs large amounts of vaccine to indirectly protect the high-risk groups by vaccinating the high-transmission ones," they wrote in a summary of their work. Tragically, the countries that need the most vaccine the soonest are the least likely to get it.

Though vaccine manufacturers like GlaxoSmithKline and Sanofi-Aventis have promised to donate millions of vaccines to developing countries, and the WHO can pressure developed nations to donate 10 percent of their stockpiles, these gestures are still woefully inadequate. After contemplating the imbalances in H1N1 vaccine distribution, Dr. Tadataka Yamada of the Bill & Melinda Gates Foundation's Global Health Program was so disturbed that he wrote, "I cannot imagine standing by and watching if, at the time of crisis, the rich live and the poor die." With the Gates Foundation, he published guidelines for the global sharing of vaccines, arguing passionately that "rich countries have a responsibility to stand in line and receive their vaccine allotments alongside poor countries."

When H1N1 spread far enough that the WHO declared it a pandemic, scientists worked rapidly to synthesize a vaccine and manufacturers churned it out. Still, the vaccine wasn't available until many months *after* the pandemic had subsided, and developing countries weren't able to afford as many doses as developed ones. Luckily, this particular strain of the flu was very mild, but the world economic situation is one major reason vaccines may not be the best weapon against pandemics.

Decentralized treatment and "hedging"

What about the most obvious strategy? That would be treating the sick with medicines that kill the pandemic disease. In the case of flu, the treatments come from a few antiviral medications. In the case of a new outbreak of the bubonic plague, we'd look to antibiotics. But we have to ask ourselves the same questions we did when considering how to use vaccines to stop a pandemic: How will we get enough medicine to enough people fast enough?

The answer isn't just to send everybody to the hospital. First of all, people may be sick in areas where there are no hospitals, and second, during a pandemic, hospitals will be overwhelmed with sick people already. Plus, sick people may not actually be able to get out of bed and go to the hospital—especially if everybody in their family is sick, too. The University of Melbourne's Robert Moss is an immunization researcher who points out that we're going to need to come up with some novel ways of delivering antivirals in the event of another pandemic. After researching the ways antivirals were prescribed during the H1N1 pandemic, Moss and his colleagues discovered that medicine wasn't handed out in a timely fashion because of one simple bottleneck: testing facilities. Most doctors conscientiously sent out blood samples from every person who visited them claiming to have the flu, and waited to hear back from often distant labs for diagnosis. As a result, people went untreated and more cases piled up as labs were overwhelmed. During a more deadly pandemic, the situation would have been disastrous.

Moss believes there are a few simple ways that doctors can simplify the process of prescribing medication to avoid this bottleneck. He calls it decentralization. If a pandemic is under way and labs are overrun, the best way to diagnose patients is based on the symptoms that they present with. Do their ailments sound like the pandemic disease? Then give them the medicine. There's no time to waste. In addition, Moss recommends setting up informal treatment centers in as many places as possible, including online, to make it easy for people to get diagnosed. Nurses who can't normally prescribe medicines should be allowed to prescribe the antivirals if a patient has the symptoms of the pandemic disease. And couriers should deliver them to people's houses.

The developing world might be better prepared for this decentralized method than the developed, mainly because many medicines in these countries are already handed out via decentralized, informal treatment facilities. Health-care workers treating everything from yellow fever to cholera have set up treatment stations in remote regions, hoping to reach the largest number of people.

The University of Hong Kong's Joseph Wu says his models show that countries should always "hedge" by stockpiling two different antiviral medications. That's because viruses often mutate during flu season, becoming resistant to the drugs used in treatment. But if we dispense two different drugs, the virus can't mutate fast enough to keep up. Wu's "hedging" strategy seems to work, at least in computer simulations of outbreaks in urban areas that assume people are traveling between cities fairly rapidly. If the city where the outbreak occurs uses two drugs to combat the pandemic instead of one, 10 percent fewer people overall will be infected than if only one drug is used. And the number of people infected with mutant strains of the virus will go down from 38 percent to 2 percent of the population. Those numbers are quite significant, especially because one of our goals as we stop a pandemic in its tracks is to prevent the microbes from mutating into something we can't treat at all.

So what would be an ideal response in the event of a pandemic? As soon as the WHO declares an outbreak, vaccines and at least two different kinds of medicines would be rushed to the most affected regions in order to stop the outbreak from becoming a pandemic. Children would be vaccinated first. If there were no vaccine available, scientists would immediately begin working on synthesizing one. Informal treatment stations would be set up in any region where there had been cases of the disease, so that people could be treated quickly, without bottlenecks. As individuals, we'd take great care to avoid big public gatherings and try to stay home as much as possible. Above all, we'd want coordination between health-care workers and pandemic modelers to figure out the best treatment strategy for each area, given the often limited resources we'll have at hand.

The main thing to remember is that stopping a pandemic isn't about treating individuals—it's about treating groups who are the most likely to spread the pandemic to others. If your neighbor's child gets a vaccine, this measure alone could protect your whole neighborhood more than if you and your adult friends all got vaccines.

Likewise, if we can kill a pandemic brewing in a developing country by sharing our vaccines and antivirals, we will save the developed world, too.

A global-health surveillance state would look very similar to the world we live in now, except there would be considerably more rapid sharing of information between groups you wouldn't expect, such as Google and the CDC, or regional Chinese hospitals, mathematicians who model pandemics, and researchers at GlaxoSmithKline. Once a region began to exhibit signs of a flu outbreak, whether reported by doctors or revealed by searches on Google, the WHO would be alerted instantly. Pandemic modelers and vaccine manufacturers could respond with strategies for containment before a pandemic virus even had a chance to jump to a new city.

Anonymous health surveillance will be an integral part of pandemic-proof cities in the future. Combine surveillance with a good system for modeling pandemics and a supply of at least two antivirals, and you've got the blueprint for one of the healthiest cities the world has ever seen. And that's the kind of city we could survive in for centuries. Unless, of course, we are dealing with a disaster that is so unusual, and so powerful, that we have no models for how it might work. All we can do in that case, as we'll see in the next chapter, is go underground.

17. CITIES THAT HIDE

THERE ARE SOME disasters so catastrophic, and so rare, that we have very little data on them. Call them extreme radiation events. As we learned in chapter two, such an event may have caused the second-worst mass extinction on Earth 450 million years ago. Some scientists speculate that a gamma-ray burst coming from a nearby hypernova may have abruptly ended the Ordovician period, frying the ozone layer off our atmosphere and exposing the Earth's newly diversified multicellular creatures to high doses of radiation. Creatures deep in the sea would have been protected by the radiation-absorbing water, but all the plants and light-loving swimmers near the surface would have been cooked instantly. Those not boiled to death would have been eaten away by radiation fairly quickly as the sun's ultraviolet rays beamed down on the unshielded planet. A gamma-ray burst like that could hit the planet at pretty much any time, with very little warning. We would likely be able to see the hypernova with our naked eyes, and would have a few short hours before its stream of radioactive particles showered down on the Earth.

Such gamma-ray bursts are a very real threat, but they're extremely unusual. Even less likely than a hypernova is the probability that its gamma-ray burst would be aimed directly at us. A far more likely cause of sudden radiation bombardment on Earth is human warfare. Even a limited nuclear war could pour ionizing radiation down on the planet in the form of nuclear fallout, causing radiation sickness in the short term and triggering cancers that could kill in the long term.

For people living in cities, it won't matter if the radiation comes from space or nuclear bombs. To survive, we'll need to go underground, into subterranean cities whose walls are made of thick layers of rock that can block radiation. We've known this for a long time. One of the greatest underground cities of the modern world, the NORAD (North American Aerospace Defense Command) complex beneath Colorado's Cheyenne Mountain, was designed during the Cold War to protect up to 5,000 residents from atomic blasts and the subsequent fallout. But this isn't the only underground city humans have built to protect themselves during war. Nearly two millennia ago in what is today central Turkey, Jews and Christians fleeing Rome built villages on top of vast underground cities that could protect thousands of people from Roman raiders—and later, from Muslims during the Crusades.

We've been surviving underground for a long time. And city planners today are building more and more underground structures. You may not think you have a chance of surviving during a radiation emergency, but you're closer to an underground city than you think.

Underground Life

The mounded hills and deep, curving valleys of Göreme, Turkey, are famous for their beauty, and for thousands of undeniably phallic rock columns known as "fairy chimneys." Though these structures elicit everything from giggles to New Agey declarations about "lingam power" from tourists, what's most interesting about them isn't their peculiarly erect shapes. Wind and water have worn them down in this way because the fairy chimneys, like the valleys and cliffs they emerged from, are made of tuff, a pale, crumbly rock composed of highly compressed volcanic ash. Early settlers in Göreme and many neighboring towns in Turkey's Cappadocia region discovered that tuff was incredibly frangible, and easy to dig out and remold. Small groups of Judeo-Christian mystics driven out of Rome in the second century CE came to central Turkey to hide, and dug spartan monk's cells into the tuff.

Over the next several hundred years, these tiny settlements of hermits and outcasts grew. Villagers carved out homes, churches, and vast food-storage pantries, eventually creating a breathtaking architectural style in which gorgeous classical columns and arched church doors appear to emerge from solid rock. Local aristocrats funded incredible subterranean art projects, deep cave churches whose high ceilings are painted with gorgeous biblical scenes that could only be seen in torchlight. Many of the most awe-inspiring examples are still preserved in Göreme's Open Air Museum, a cluster of churches and monasteries carved into the creamy tuff above a valley full of patchwork farms. From a distance, the ornate, sculpted complex looks like an otherworldly city from a *Lord of the Rings* movie. During the Byzantine era, however, the need for these cave and tunnel cities was all too real.

Cappadocians built dozens of underground cities in the era between the fifth and tenth centuries. Historians have conflicting theories about why, but one major reason would no doubt have been raids from neighboring groups and from Muslims during the Crusades. These underground cities don't just shield inhabitants from the elements–their entrances are hard to reach, or hidden. They are designed to be invisible.

To imagine how future humans might survive a radiation disaster underground, we need only pay a visit to subterranean Cappadocia. On a drizzly day spent near Göreme, I ducked out of the summer rain to visit Derinkuyu, the most extensive of the excavated underground cities in the region. Now open for tourists, the city's maze of tunnels, living quarters, and community areas extends five levels and 55 meters underground. Cool and sandy, its often cramped corridors are well ventilated by several air shafts. My tour group and I squeezed down a long stairway, then entered a large room that was once a stable. It could easily have housed a dozen goats, sheep, or cows, chomping contentedly from mangers carved right into the walls. Our guide pointed out that villagers typically entered Derinkuyu via secret tunnels from their aboveground homes. These entrance tunnels were small enough in many places that I was forced to bend quite far down as I walked, and I'm a fairly short person. The city's builders engineered these areas to discourage anyone trying to enter with bulky weapons or armor. Punctuating many of the stairways were deep crevices that held enormous rock discs designed to roll out and block the corridor from invaders.

Inside the underground city of Derinkuyu, you can see two hallways leading to rooms. The city is five floors deep and housed thousands of people.

Living quarters were honeycombs of interconnected rooms and bed nooks hollowed out of the tuff walls. Though the place looked barren when I saw it, a thousand years ago it would have been very different. People built wooden doors into the round doorways, covering the floors in thick carpets and the walls in draperies. Families from the city above had their own quarters below, full of furniture, favorite pottery, food, and wine. In the living areas of the city, ceilings were high and rooms were cozy rather than cramped. The underground dwellers even had a sanitation solution for long-term sojourns. Waste was packed into clay containers, sealed up, and buried in deep pits below the city's lowest level.

Large public rooms for cooking and eating, as well as wine-making and worship, would have been places where villagers could gather to make decisions about how long to remain underground and hide from danger.

This ancient city, carved out over centuries, was constantly changing and growing. A deep passage connected Derinkuyu with another underground city several kilometers away. The peoples who made these passages their refuges were employing the same survival mechanisms used by Jews over the past 2,000 years—they had scattered from distant communities in both the west and the east, and adapted themselves to the remote landscapes of central Turkey to protect themselves from persecution and ethnic cleansing. Not only did these communities survive for centuries, but they also created an entirely new way of life—one that many people in the area still enjoy today. Traditional villages in Cappadocia are full of homes built into the tuff of fairy chimneys, with narrow stairways spiraling around the rocks' girth, leading to sturdy wooden doorways in their cone-shaped tops. Some homes are hewn from the cliffs, where they share space with countless pigeon roosts that locals tend for the fertilizer. Tourists are invited to spend the night in refurbished cave homes, as bed-and-breakfasts take over abandoned dwellings. I spent several days in one such hotel, my bed stashed deep inside a cave that had been modified to have large picture windows overlooking the city of Göreme. With the exception of the windows, my bedroom could have been torn from a future world where humans had to relocate underground for protection.

But without those windows, the place would have been much more dismal. And that's why urban planners creating modern underground cities worry as much about the psychological effects of living underground as they do about the structural integrity of underground spaces.

The Trouble with Tunnels

If we're going to survive a nuclear war, a meteor strike, or a radiation event, there is no doubt that we'll have to live underground for months, or even years, as the planet recovers. In the unlikely event that a gamma-ray burst burns off the ozone layer, it could be centuries before life could thrive on the surface again—whenever we stepped outside our underground homes, we'd need to wear protective covering to shield us from the sun's ultraviolet radiation. The good news is that three feet of packed dirt over your head can significantly reduce the intensity of radiation, and a layer of concrete can provide more safety still. We have the engineering ability to create radiation-shielded cities by going underground. The question is how we would live there.

In the event of a radiation emergency, people might find themselves having to create cities in already-existing underground spaces like subway tunnels, mines, sewer systems, and service tunnels. Already-existing underground cities like Montréal's "RÉSO," a 20-mile system of tunnels that connect shopping centers, metro stations, schools, apartments, and more, are basically larger and more complicated versions of mining shafts. To make the long, dark corridors more livable, developers build structures inside the bare rock walls, covering up the stone surfaces.

Making these spaces inviting is crucial to our survival. John Zacharias, a city-planning professor at Montréal's Concordia University, has studied several underground cities, especially in Japan and China, and told me that the biggest challenge is psychological. Studies on people who work all day in underground space without any access to the outside show rising stress levels. "It's not dramatic, but is measurable," he said. "Going down very deep is also something people don't like." The new Oedo subway line in Tokyo is 55 meters below the surface, and Zacharias said people tend to avoid it in favor of an overcrowded line it was supposed to relieve. In Finland and Sweden, where underground buildings are common, studies have shown that people are disturbed by the process of descending into the Earth, and that they complain of the monotony in subterranean buildings. The solution, argue civil engineers John Carmody and Raymond Sterling in *Underground Space Design,* their underground-engineer omnibus, is to make sure underground spaces are "stimulating, varied environments" that give the impression of spaciousness and daylight.

Many underground cities, like RÉSO, use skylights to bring in daylight, but our future troglodytes won't have that option because they will need radiation shielding. So they'll have to arrange underground areas to be different sizes and shapes, with architectural features that make the space interesting to inhabit. Even the residents at Derinkuyu knew this, and their interior spaces were all uniquely arranged, with a wide variety of floor plans. Carmody and Sterling also caution that one of the main complaints people have underground is that they become disoriented without windows, so a good underground city would need a simple layout or clear signs that help inhabitants find their way. Because people get anxious about going deep underground, transitions between levels should be gradual. Ideally, different areas of the city would have dramatically different designs to give the feeling of neighborhoods and landmarks that we use aboveground to figure out where we are. Privacy will also be a premium in spaces like these, where people have a tendency to feel trapped. As we turn our tunnels into our homes, we'll want to remember to create places to be alone, as well as vast, high-ceilinged rooms that will make us feel as if we're outdoors even if we aren't.

If we are turning mines into cities, or excavating a brand-new subterranean metropolis, there are also a few basic engineering issues we should keep in mind. Agust Gudmundsson, a geology professor at Royal Holloway, University of London, studies underground structures, and he explained that earthquakes will be a threat to life belowground. "Fractures that form or are reactivated during an earthquake may lead to water flowing into parts of the underground city," he said. Water leakage is one of the main causes of destruction in these developments, and leaks recently led to partial collapse of some tunnels in RÉSO. When we build underground, Gudmundsson cautioned, we have to be vigilant about whether we're building in regions with faults or cracks—especially if the city will be near a body of water. Just building the city could cause tremors that allow water to come seeping through cracks, causing damage or collapse over time. At the NORAD facility in Colorado, the water-leakage problem was so severe that people walked under umbrellas through the long rock tunnel that winds deep into the mountain and to the massively reinforced city gates.

Nowhere to Go but Down

"The construction of major transit projects such as metros and road tunnels is just a prelude for the true nature of underground development," argue Dmitris Kaliampakos and Andreas Barnardos, two engineers who specialize in underground development at the National Technical University of Athens. In 2008, the two helped organize an international conference to deal with questions about building underground. Over the past few decades, many cities have witnessed the proliferation of underground building projects, ranging from individual homes expanding underground to RÉSO-like structures such as the one Amsterdam considered building beneath its canals in the late 2000s. "The main driving force behind the process is the continuously growing urban areas," write Kaliampakos and Barnardos. Building underground saves energy, they point out, because subsurface temperatures stay comfortable year-round; in addition, it allows cities to expand without destroying historic places or producing suburban sprawl. The problem is that most cities don't have a lot of regulations and codes in place to help developers make these underground spaces. It's hard for real-estate agents to value underground spaces that don't exist, and most laws don't make it clear who owns the land below our feet. These factors, along with the cost of building down, make developers shy about breaking ground.

Still, the laws are catching up with urban needs. And geologists like Gudmundsson are working with engineers to provide accurate maps of the kinds of rocks and fissures that lurk beneath us, so that we can plan the right spots to tunnel below. As John Zacharias said, "We are going to have a lot more space underground in future, especially as cities build new transportation systems underground." He predicted that the movements below will also be related to concerns about energy. "We will need places to store water, especially as cities get round to recycling water," he asserted. "Power plants will go underground. Theaters and libraries are already there. The future city is three-dimensional, and all big cities will be looking to see how they can better use the underground resource."

As more cities send vital roots underground, we create a world that is inadvertently preparing itself for a radiation emergency. The more we make the subsurface livable, the more likely it is that humans will survive to see the next several millennia.

In this description of underground cities, we've considered city designs that would make us comfortable living underground, and we've learned that our worst enemy underground will be seeping water. But we've danced around the real issue we'll confront in our radiation-proof cities: food. As the atmospheric scientist Alan Robock of Rutgers University points out in one of his many papers on nuclear winter, the biggest issue we'll face may not be radiation at all. It will be starvation in the wake of extensive burning:

> Smoke—especially black, sooty smoke from cities and industrial plants—would block sunlight for weeks or months over most of the Northern Hemisphere. And, if a nuclear holocaust occurred in the Northern Hemisphere in summer, it would affect much of the Southern Hemisphere as well. The cool, dark conditions at the earth's surface would eliminate at least one growing season, resulting in a global famine.

Famine will also be a problem if one of our planet's many megavolcanoes goes off. Ash and soot from such an enormous eruption would be blasted into the stratosphere, cutting the planet off from life-giving sunlight. The atmosphere would likely be full of sulfides and dust as well, both of which we'd want to avoid. So we might be looking at generations who live much of their lives underground. Our underground cities will have to be farms as well as shelters. In the next chapter, we'll explore in greater detail how such farms might work.

Just like underground cities, farm cities are being built already, for many of the same reasons. Farm cities, like the green cities the urban geographer Richard Walker described in his book on San Francisco, are far more energy efficient and environmentally sustainable than the industrial cities most of us inhabit today. They are also less likely to suffer famine. In the next chapter, we'll speculate about what cities might look like in a century or two. It's possible they'll be nearly indistinguishable from the natural surface of the planet.

18. EVERY SURFACE A FARM

WE'VE EXPLORED HOW cities are not static objects to be feared or admired, but are instead a living process that residents are changing all the time. Given how much bigger and more common cities are likely to become in the next hundred years, we'll need to change them even further. Using predictive models from the fields of engineering and public health, our future city designers will plan safer, healthier cities that could allow us to survive natural disasters, pandemics, and even a radiation calamity that drives us underground. But there is a yet more radical way we'll transform our cities. Over the next two centuries, we'll probably convert urban spaces into biological organisms. By doing this, we make ourselves ready to prevent two of the biggest threats to human existence: starvation and environmental destruction.

Eventually this biological transformation might result in cities unlike any that have existed before. But for now, the best way to understand how such a shift would begin is by paying a visit to a city park or garden. These are places that we've built in the middle of cities to closely resemble the natural world. Usually they are just as engineered and artificial as the buildings surrounding them, but they do a lot of things that buildings typically can't do, such as sequester carbon, absorb runoff storm water, and provide a cool, shady environment without drawing any energy from the grid. Many city parks today are reclamations of previously blighted areas. In Vancouver, Canada, for example, residents of the Fairview neighborhood converted a stretch of abandoned railroad tracks into dozens of garden plots where locals grow vegetables, flowers, and grains around the still-visible iron rails. And in New York City, a group of enthusiasts lobbied the city to let them convert a historic elevated-train structure into a park, which is now called (appropriately enough) the High Line. This once abandoned viaduct now features trees and grasses that seem to sprout from its concrete columns. People in these cities and many others throughout the world are slowly blanketing their barren causeways in habitats where plants and animals can thrive.

If we want the populations of our cities to survive, however, we're going to have to do a lot more than plant flowers in lower Manhattan. We'll need to transform urban areas into regions that can, as much as possible, feed themselves. That means prairie cities can't rely on distant countries for bananas, nor can people living in desert outposts expect to get grain from fertile basins hundreds of kilometers away. More pressingly, we need to build cities that draw energy from their local ecosystems. By growing biofuels, and using sunlight for power, we make it less likely that humanity's home planet will one day no longer sustain our need for energy. The biological city could provide us with food and energy security for millennia to come.

Food on the Streets

When I visited Cuba in the early 2000s, the best places to buy fresh food in Havana were street markets where urban farmers sold whatever they'd cultivated on roofs or in window boxes, sidewalk gardens, and yards. I wandered around in one of these markets, located in a large, airy warehouse where a couple of dozen people had set out their goods in baskets and on blankets. One woman was selling four eggs, a few eggplants, and a cellophane bag of spices. Another sat back on her heels behind a blanket heaped with greens. Street markets occupied a precarious legal posi-

The Eden Project in Cornwall is an experiment with environmentally sustainable architecture; each dome contains its own ecosystem. In the future, cities might grow food or energy sources in such domes, or they might serve as water- and air-filtration devices for eco-buildings.

In this example of eco-architecture, a hotel's living walls are fed by a rooftop water source. The photograph was taken by Robinson Esparza, in the Huilo Huilo preservation area in Chile.

tion under communism because they encouraged private enterprise. But instead of cracking down, the Cuban government was paying agricultural engineers to study the most productive methods of urban farming. The need to prevent starvation overrode ideological concerns.

Though that ad-hoc urban farmer's market in Havana felt like a medieval oasis in the middle of a bustling, cosmopolitan city, it was actually a good demonstration of how people might grow and buy their food in the cities of tomorrow. They'll do it by slowly converting cities into farms. At the time I was in Cuba, Raquel Pinderhughes, an urban planning professor at San Francisco State, wrote that there were over 8,000 farms in Havana, covering about 30 percent of the region's available land. If you rode into the countryside on a bus that picked you up on Havana's busy Malecón, a promenade along the seawall, you'd find that the high-density city quickly shaded into suburban residential areas peppered with farmland. Land planners sometimes call this system periurban agriculture. It transforms suburban consumer sprawl into a rich source of food production.

In the hot, dry valleys of Pomona, California, a nonprofit group called Uncommon Good has helped set up an urban farm where unemployed immigrants with farming experience grow organic food to sell in local markets. This Pomona farm, like many others, uses the "small-plot intensive farming" (SPIN) model, designed by urban farmers in Canada to maximize crop yields in areas of less than an acre. The idea behind SPIN is both agricultural and economic. Farmers vary their crops and use sustainable fertilizers to keep their small plots of soil fecund, and they sell by direct marketing in their local areas. This maximizes food production and minimizes the resources that the farmers need to transport that food to buyers. It's easy to imagine many cities transformed by a SPIN model over the next 50 years, where people grow their own food to eat and sell to neighbors—who in turn sell different food, so that local diets can remain varied.

But will cities transform farming as much as urban farmers hope their methods will transform cities? In his book *The Vertical Farm: Feeding the World in the 21st Century*, Columbia University environmental-health professor Dickson Despommier argues that cities of the future might feed themselves by creating farms inside enormous, glass-walled skyscrapers where every floor is a solar-powered greenhouse. All water in these skyscraper farms would be recycled, and the structures themselves would be designed to be carbon neutral. While critics question whether it would be possible to heat, power, light, and tend skyscraper farms without wasting a lot of energy, Despommier's thought experiment is a good one. We are going to need ways to produce enormous amounts of food in cities, often indoors, and trying to figure out how we'd do that in a skyscraper—or an underground cavern, for that matter—is a step in the right direction.

Our future buildings may be sprouting gardens on the outside, too. A popular way to transform cities in Germany is by building green roofs, which are basically special systems designed to convert rooftops into gardens. This isn't just a matter of heaping some dirt up and throwing seeds on it. Green roofs are a complex system of layers designed to protect the roof, absorb water, and hold soil in place. Though they are unlikely to be useful for farming, some studies have shown that green roofs help cut energy costs by keeping buildings cooler in the summer months. They also reduce storm-water runoff, which is a huge issue in cities. Because most cities are covered in nonporous, nonabsorbent surfaces, all the grime, toxins, and trash in the city are washed out by rainwater during storms—and carried into nearby waterways, farms, and oceans. Having a roof that can absorb rainwater does a tremendous amount of good for the local environment and cuts costs related to water purification and treatment.

Bringing natural environments into cities isn't just about feeding ourselves. It's also about figuring out how to manage our energy consumption using tricks borrowed from nature. Growing shade plants on our roofs can help cut energy costs in summer, just as designing photosynthetic antennae like the ones mentioned in chapter 11 can help us power computers without burning coal. Natural ecosystems conserve energy remarkably well. As we learn to imitate that, our cities might become highly advanced technological entities that look strangely like the post-apocalyptic jungle version of New York City in *The World Without Us*.

Managing the Land

MIT's environmental-policy professor Judith Layzer offered me a vivid picture of what life might be like in such a city. She believes that, ideally, most future human communities would be based in cities, leaving enormous stretches of land free for farms and wildlife. "We need to re-regionalize," she told me. "A global economy doesn't make sense environmentally. So your ecosystem would become your bioregion." She described a world where communities would be organized around bioregions like the dense forests and rocky coasts of the mid-Atlantic states or the prairie grasslands of North America's Great Basin. "Most of your food should come from your region," she said, and farm labor would be done by people rather than machines. But, she asserted, "nobody would be working as hard as they are now" because life would have a much slower pace. "You'd have goats mowing lawns," she said, her face quirking into a grin. "It would be less efficient in the contemporary sense. Long-distance travel would be more of a hassle. You'd bike everywhere." In her ideal city, where food was local and energy carbon neutral, "you'd do everything with natural systems." And the population of a city would never rise above a few million.

This kind of regionalism might be good for our ecosystems, but it probably couldn't be as "natural" as Layzer imagines. Obviously, if people are depending on their bioregion for food, they'll be more vulnerable to the vicissitudes of climate and seasonal drought. We'll need cutting-edge technology to help these bioregional cities weather periods when the local ecosystem can't support the population.

One possible way we'll do this is by looking to outer space. At UC Santa Barbara, an international group of climate scientists, geographers, and geologists use satellite data to predict where drought will strike next. They call themselves the Climate Hazards Group, and their success at predicting drought is almost uncanny. Currently, they focus most of their efforts on Africa. Amy McNally, a geography researcher with the group, said she and her colleagues helped predict the summer 2011 drought in Somalia by correlating data drawn from satellite images of rainfall in the region with rain gauges on the ground. "They predicted the drought and the resulting famine a year in advance," she said. Unfortunately, "even with that much forewarning, response didn't make it in time for it to not get to a famine-level crisis." But the group had gained more evidence about what signs indicated droughts to come.

One of the key indicators comes from satellite observations of greenery on the ground. Just as green roofs keep buildings cooler, a green ground cover keeps the soil cooler, wetter, and more likely to yield a good crop. When plants die back too much, drought may be on its way in the next season. McNally said the satellite she uses measures the wavelengths of light reflected back from the West African region she studies. Plants reflect green light back into space, where the satellites measure percentages of green light versus other wavelengths. As a result, McNally can get an extremely precise picture of how much green is required on the ground to guarantee a good growing season. The big issue in Africa is that most regions don't use extensive irrigation, so farmers are dependent on rainfall for a successful crop. A dry season can mean death. But it doesn't have to.

Knowing we've got an impending drought might mean shoring up water supplies for irrigation that could keep a valuable plant cover protecting the soil. As our cities become more closely tied to their bioregions, science teams like the Climate Hazards Group could become crucial to urban planning. With the technology and data we have now, McNally said, "we can make predictions like 'In the next twenty years, you'll have five droughts, which is two more than usual.' " This kind of information could prove invaluable to farmers planning their water usage, or governments trying to set up trade arrangements with areas that won't be affected by the drought. As we gather more data on how droughts happen, we may be able to make more accurate predictions about when famine is likely to strike—and stop it before it starts.

Satellite imagery and technology are not a panacea for food-security problems. In fact, as we discussed earlier, famine is usually caused by political and social upheaval. Fixing that will require more than good science. You could say the same thing about our energy problems. But it's possible that our political priorities will change along with our changing urban environments.

The Biological City

As we move further into the future, our cities won't just be swaddled in gardens and farms. They might also become biological entities, walls hung with curtains of algae that glow at night while sequestering carbon, and floors made from tweaked cellular material that strengthens like bones as we walk on it. New York architect David Benjamin is part of a new generation of urban designers who collaborate with biologists to create building materials of the future. I met him in Studio-X, a branch of Columbia University's school of architecture located in a bare-bones whitewashed work space south of Greenwich Village. Students focused on monitors full of three-dimensional renderings of buildings, or sketched at drafting tables between concrete columns. It looked like the kind of place that could, in 50 years, be sprouting a layer of grass from its walls—or something much stranger than that.

Benjamin described the shift to biological cities using quick, precise gestures that reminded me of someone penciling lines on a blueprint. "It might look the way it looks now," he said. "The city could be made with bioplastics instead of petroleum plastics, but it would look very similar. A machine for making genetically modified organisms (GMOs) would exist in factories the way they do today for making medicine and biofuels." So the plastic fittings around windows would be manufactured from modified bacteria rather than fossil fuels, but as a city dweller you'd notice little difference. Benjamin and a group of other architects and biologists have worked with Autodesk, the company that makes the popular AutoCAD software many architects use to design buildings, to create a mock-up of AutoCAD for biological designs, called BioCAD. Pulling out his laptop, Benjamin showed me a demonstration of the biological-design-software interface. The designer can choose between biological materials with different properties, like flexibility or strength. Having chosen those, the designer directs the program to create structures that look like marble cake, a multicolored swirl of substances combined into a single structure that gives in the right places and holds steady in others.

Over time, these living cities would start to look different. They'd be transformed by synthetic biology, a young field of engineering that crafts building materials from DNA and cells rather than more traditional biological materials like trees. Benjamin described a recently created synthetic-biology product called BacillaFilla, designed by a group of college students in England. The students engineered a common strain of bacteria to extrude a combination of glue and calcium when put into contact with concrete. They applied the bacterial goo to cracks in concrete, and over time it filled the cracks completely and then died, leaving behind a strong, fibrous substance that has the same strength as concrete. The students described BacillaFilla as the first step toward "self-healing concrete," and their efforts are just one among many designed to create biological substances that could heal ship hulls, metal girders, and more.

Extrapolating from this development in synthetic biology, Benjamin mused, "Maybe you could program a seed to grow into a house. Or maybe cities would be so in tune with ecosystems that they would grow over time, and then decay over time, too." Synthetic biology might also help solve one of the biggest problems with new buildings, which is water leakage. Architects could design a building that is semi-permeable, with membranes that allow the circulation of air and water at various times. It's easy to imagine a future architect fashioning just such a thing with BioCAD, with patches of permeable materials built right into the fabric of the walls. The water could be purified and used, and the air would become part of a natural cooling or heating system. This building might also use computer networks to monitor its community of local buildings to figure out when to gather solar energy and send it to the grid to share, and when to lower louvers to keep residents cool. "I sometimes imagine urban landscapes that are integrated into their ecosystems with a combination of vegetation and constructed materials," Benjamin said. "They look almost like ruins in the jungle but they're actually fully functional, occupied cities."

Benjamin's visions of the future end where his fellow synthetic-biology designer Rachel Armstrong's begin. Armstrong, who is based in London, is an outspoken advocate for what she calls "the living city," or urban structures that she told me we'd create in the same way we cook or garden. We met in a café in the heart of London, overlooking the busy Tottenham Court Road, and almost immediately Armstrong was imagining how she'd rebuild the city around us. "We'd have biofuel-generated façades, or technology based on algae," she said, pointing at the windows. "You'd have surfaces creeping down buildings like icing. Strange, colored panels would glow through windows at night, and you'd have bioluminescent streetlights. Bridges will light up when we step on them." She paused, but continued staring outside, deep in thought. "We'd keep the bones of buildings steel and concrete, but rewrap those spaces with increasingly more biological façades. Some will be porous and attract water; others will process human waste. Mold won't be something you clean off a surface but will be something you garden."

Armstrong is fascinated by bacteria and mold, which she and other synthetic-biology designers view as the building blocks of future cities. "We are full of microbes," she asserted firmly. "Maybe instead of using environmental poisons to create healthier environments inside, we should be using probiotics." Glowing bacteria could live in our ceilings, lighting up as the sun goes down. Other bacteria might purify the air, scrubbing out carbon. Every future urban home would be equipped with algae bioreactors for both fuel and food.

Her vision isn't just science fiction. Recently, Armstrong worked with a group of biologists and designers who hope to use experimental proto-cells—basically, a few chemicals wrapped in a membrane—in a project that could prevent Venice from sinking into the water. Proto-cells are semi-biological, and can be designed to carry out very simple chemical processes. In Venice, engineers would release proto-cells into the water. Designed to prefer darkness, the proto-cells would quickly head for the rotting pilings beneath the city's dwellings. Once attached to the wood, the cells would slowly undergo a chemical transformation in which their flexible membranes transformed into calcium shells. These calcium shells would form the core of a new, artificial reef. As wildlife discovered the calcium deposits, a natural reef would form. Over time, the city's shaky foundations would become a stable reef ecosystem. Already, Armstrong and her group have had some success creating small-scale versions of the proto-cell reef in the lab, and they're moving on to experiments in controlled natural areas.

If our cities do evolve to be more like biological organisms and ecosystems, it could change the way communities form within their walls. "We might start to experience the city as something we have to take care of the way we take care of our bodies," Armstrong suggested. In a biological city, using toxic chemicals in your kitchen might cause your algae lights to die. "We'll take more care of the city because we feel its injuries more deeply," Armstrong said. It's possible that this would generate a sense of collective responsibility for our buildings and avenues. Neighbors would tend their buildings together, trading recipes for making fuel the way people today trade recipes for holiday cakes.

Armstrong's hyper-technological biometropolis shares something in common with Judith Layzer's vision of small, slow cities devoted to farming. Both of them arise from the belief that city dwellers will become producers rather than consumers. With home bioreactors, Armstrong said, "our spaces will become a place where we can generate wealth." This idea is central to the SPIN model of urban farming, too. "It's about decentralizing energy and food production, basically," Armstrong concluded. Of course, it's impossible to predict what the consequences would be for people in cities whose buildings were half-alive. Armstrong is willing to admit her ideas are utopian, and that's the point. "You need something to aim at," she said with a smile.

As we look further to the future in the next part, we'll be taking aim at something even more speculative than self-healing cities that look like glowing ruins and sprout food and power from every surface. We'll see what humanity might become if we manage to survive for another million years. To do this we're going to need to do more than rebuild our cities. We'll need to rebuild the entire Earth. And then we'll start striking out for the territories beyond our planet, lifting ourselves into space and colonizing the solar system. What will humans be like after tens of thousands of years of evolution, especially in space? It's possible that our progeny will be as unlike us as we are from *Australopithecus*—and yet they will still be as human as our distant hominin ancestors were.

PART V THE MILLION-YEAR VIEW

19. TERRAFORMING EARTH

THE LONG-TERM GOAL for *Homo sapiens* as a species right now should be to survive for at least another million years. It's not much to ask. As we know, a few species have survived for billions of years, and many have survived for tens of millions. Our ancient ancestors started exploring the world beyond Africa over a million years ago, and so it seems fitting to pick the next million years as the first distant horizon where we'll set our sights. We've already talked about how we can start this journey by building cities that are safer and more sustainable. Eventually, however, we're going to build cities on the Moon and other planets. Our future is among the stars, as the science-fiction author Octavia Butler suggests. But long before we have the technologies to get there, our survival will depend on looking at Earth from the perspective of extraterrestrials.

Imagine we're on an interstellar voyage and we encounter an Earth-like planet. As we survey it from orbit, we discover that this planet is full of life, and covered with sprawling artificial structures built by a scientifically advanced civilization. Seeing those, most of us would say the planet is controlled by a group of intelligent beings. That's the extraterrestrial perspective. Right now, we're stuck in the terrestrial perspective, where we do not really regard ourselves as "controlling" the Earth. Nor do we see ourselves as one group. From space we might look like a unified civilization of clever monkeys who hang out together building towers, but down on Earth we are Russians, Nigerians, Brazilians, and many other identities that divide us. Our differences aren't always a problem. But they have so far prevented us from coming up with a global solution to maintaining the Earth's resources. We won't make it into the far future unless we start banding together as a species to control the Earth in a way we never have before.

When I say "control the Earth," I don't mean that we'll all shake our fists at the sky and declare ourselves masters of everything. As entertaining as that would be, I'm talking about something a bit less grandiose. We simply need to take responsibility for something that's been true for centuries: Human beings control what happens to most ecosystems on the planet. We're an invasive species, and we've turned wild prairies into farms, deserts into cities, and oceans into shipping lanes studded with oil wells. There is also overwhelming evidence that our habit of burning fossil fuels has changed the molecular composition of the air we breathe, pushing us in the direction of a greenhouse planet. More than at any other time in history, humans control the environment. Still, our environment is going to change disastrously at some point, whether it's heated by our carbon emissions from fossil fuels or cooled by megavolcano eruptions.

Our first priority in the near future must be to control our carbon output. I cannot emphasize this enough. As environmental writers like Bill McKibben and Mark Hertsgaard have argued, our fossil-fuel emissions are heating up the planet, and we can prevent this situation from becoming worse by using green sources of energy. Maggie Koerth-Baker points out in *Before the Lights Go Out: Conquering the Energy Crisis Before It Conquers Us* that we already have several types of sustainable energy to choose from, including solar and wind. Government representatives who attend the annual U.N. Climate Change Conferences are also coming up with strategies to encourage countries to curb fossil-fuel use, proposing everything from carbon taxes to emissions regulations.

The problem is that our climate has already been permanently changed for the next millennium, as the geobiologist Roger Summons explained in part one of this book. To prevent the planet from becoming uninhabitable, we'll have to take our control of the environment a step further and become geoengineers, or people who use technology to shape geological processes. Though "geoengineering" is the proper term here, I used the word "terraforming" in the title of this chapter because it refers to making other planets more comfortable for humans. Earth has been many planets over its history. As geoengineers, we aren't going to "heal" the Earth, or return it to a prehuman "state of nature." That would mean submitting ourselves to the vicissitudes of the planet's carbon cycles, which have already caused several mass extinctions. What we need to do is actually quite unnatural: we must prevent the Earth from going through its periodic transformation into a greenhouse that is inhospitable to humans and the food webs where we evolved. Put another way, we need to adapt the planet to suit humanity.

Over the coming centuries, we'll need to take measures more drastic than cutting back on fossil-fuel use and ramping up the deployment of alternative energies. Eventually we'll have to "hack the planet," as they say in science-fiction movies. And we'll do that in part by re-creating great planetary disasters from history.

Blocking the Sun

To make Earth more human-friendly, our geoengineering projects will need to cool the planet down and remove carbon dioxide from the atmosphere. These projects fall into two categories. The first, called solar management, would reduce the sunlight that warms the planet. The second, called carbon-dioxide removal, does exactly what it sounds like. Futurist Jamais Cascio, author of *Hacking the Earth: Understanding the Consequences of Geoengineering*, predicts that we'll see an attempt to initiate a major geoengineering project in the next 10 years, and it will probably be solar management. "It's a faster effect and tends to be relatively cheap, and some estimates are in the billion-dollar range," he said. "It's cheap enough that a small country or a rich guy with a hero complex who wants to save the world could do it." Indeed, one solar-management project is already under way, albeit inadvertently. Evidence suggests that sulfur-laced aerosol exhaust emitted by cargo ships on the ocean changes the structure of high clouds, making them more reflective and possibly cooling temperatures over the water. Some solar-management plans take note of this discovery, and propose that we fill the oceans with ships spraying aerosols high into the air. But other strategies are more radical.

To find out how we'd shield our planet from sunlight, I visited the University of Oxford, winding my way through the city's maze of pale gold spires and stone alleys to find an enclave of would-be geoengineers. Few deliberate geoengineering projects have been tried to date, but the mandate of the future-focused Oxford Martin School is to tackle scientific problems that will become important over the next century. The center is helping invent a field of science that doesn't properly exist yet—but that will soon become critically important. One of its researchers is Simon Driscoll, a young geophysicist who divides his time between studying historic volcanic eruptions and figuring out how geoengineers could duplicate the effects of a volcano in the Earth's atmosphere without actually blowing anything up.

Driscoll told me what volcanoes do to the atmosphere while cobbling together cups of tea in the cluttered atmospheric-physics department kitchen. Along with all the flaming lava, they emit tiny airborne particles called aerosols, which are trapped by the Earth's atmosphere. He cupped his hands into a half sphere over the steam erupting from our mugs of tea, pantomiming the layers of Earth's atmosphere trapping aerosols. Soot, sulfuric acid mixed with water, and other particles erupt from the volcano, shoot far above the breathable part of our atmosphere but remain hanging somewhere above the clouds, scattering solar radiation back into space. With less sunlight hitting the Earth's surface, the climate cools. This is exactly what happened after the famous eruption of Krakatoa in the late 19th century. The eruption was so enormous that it sent sulfur-laced particles high into the stratosphere, a layer of atmosphere that sits between ten and forty-eight kilometers above the planet, where they reflected enough sunlight to lower global temperatures by 2.2 degrees Fahrenheit on average. The particles altered weather patterns for several years.

Driscoll drew a model of the upper atmosphere on a whiteboard. "Here's the troposphere," he said, drawing an arc. Above that he drew another arc for the tropopause, which sits between the troposphere and the next arc, the stratosphere. Most planes fly roughly in the upper troposphere, occasionally entering the stratosphere. To cool the planet, Driscoll explained, we'd want to inject reflective particles into the stratosphere, because it's too high for rain to wash them out. These particles might remain floating in the stratosphere for up to two years, reflecting the light and preventing the sun from heating up the lower levels of the atmosphere, where we live. Driscoll's passion is in creating computer models of how the climate has responded to past eruptions. He then uses those models to predict the outcomes of geoengineering projects.

The Harvard physicist and public-policy professor David Keith has suggested that we could engineer particles into tiny, thin discs with "self-levitating" properties that could help them remain in the stratosphere for over twenty years. "There's a lot of talk about 'particle X,' or the optimal particle," Driscoll said. "You want something that scatters light without absorbing it." He added that some scientists have suggested using soot, a common volcanic by-product, because it could be self-levitating. The problem is that data from previous volcanic eruptions shows that soot absorbs low-wavelength light, which causes unexpected atmospheric effects. If past eruptions like Krakatoa are any indication—and they should be—massive soot injections would cool most of the planet, but changes in stratospheric winds would mean that the area over Eurasia's valuable farmlands would get hotter. So the unintended consequences could actually make food security much worse.

It's not clear how we'd accomplish the monumental task of injecting the particles, but Driscoll's colleagues at Oxford believe we could release them from spigots attached to enormous weather balloons. Weather balloons typically fly in the stratosphere, and they could release reflective particles as a kind of cloud while remaining tethered to an ocean vessel. The stratosphere's intense winds would carry the particles all around the globe. However, getting particles into the atmosphere isn't the tough part.

The real issues, for Driscoll and his colleagues, are the unintended consequences of doping our atmosphere with substances normally unleashed during horrific catastrophes. Rutgers atmospheric scientist Alan Robock has run a number of computer simulations of the sulfate-particle injection process, and warns that it could destroy familiar weather patterns, erode the ozone layer, and hasten the process of ocean acidification, a major cause of extinctions.

"I think a lot about the doomsday things that might happen," Driscoll said. Unintended warming and acidification are two possibilities, but geoengineering could also "shut down monsoons," he speculated. There are limits to what we can predict, however. We've never done anything like this before.

If the planet starts heating up rapidly, and droughts are causing mass death, it's very possible that we'll become desperate enough to try solar management. The planet would rapidly cool a few degrees, and give crops a chance to thrive again. What will it be like to live through a geoengineering project like that? "People say we'll have white skies–blue skies will be a thing of the past," Cascio said. Plus, solar management is only "a tourniquet," he warned. The greater injury would still need treating. We might cut the heat, but we'd still be coping with elevated levels of carbon in our atmosphere, interacting with sunlight to raise temperatures. When the reflective particles precipitated out of the stratosphere the planet would once again undergo rapid, intense heating. "You could make things significantly worse if you're not pulling carbon down at the same time," Cascio said. That's why we need a way of removing carbon from the atmosphere while we're blocking the sun.

Turning Earth into a Carbon-Eating Machine

One of the only geoengineering efforts ever tried was aimed at pulling carbon out of the atmosphere using one of the Earth's most adaptable organisms: algae called diatoms. Researchers have suggested that we could scrub the atmosphere by re-creating the conditions that created our oxygen-rich atmosphere in the first place. In several experiments, geoengineers fertilized patches of the Southern Sea with powdered iron, creating a feast for local algae. This resulted in enormous algae blooms. The scientists' hope was that the single-celled organisms could pull carbon out of the air as part of their natural life cycle, sequestering the unwanted molecules in their bodies and releasing oxygen in its place. As the algae died, they would fall to the ocean bottom, taking the carbon with them. During many of the experiments, however, the diatoms released carbon back into the atmosphere when they died instead of transporting it into the deep ocean. Still, a few experiments suggested that carbon-saturated algae *can* sink to the ocean floor under the right conditions. More recently, an entrepreneur conducted a rogue geoengineering project of this type off the coast of Canada. The diatoms bloomed, but the jury is still out on whether ocean fertilization is a viable option.

So it's possible that algae will be helping us in our geoengineering projects. Another possibility is that we'll be enlisting the aid of rocks. One of the most intriguing theories about how we'd manipulate the Earth into pulling down carbon was dreamed up by Tim Kruger, who heads the Oxford Martin School's geoengineering efforts. I met with him across campus from Driscoll's office, in an enormous stone building once called the Indian Institute and devoted to training British civil servants for jobs in India. It was erected at the height of British imperialism, long before anyone imagined that burning coal might change the planet as profoundly as colonialism did.

Kruger is a slight, blond man who leans forward earnestly when he talks. "I've looked at heating limestone to generate lime that you could add to seawater," he explained in the same tone another person might use to describe a new recipe for cake. Of course, Kruger's cake is very dangerous—though it might just save the world. "When you add lime to seawater, it absorbs almost twice as much carbon dioxide as before," he continued. Once all that extra carbon was locked into the ocean, it would slowly cycle into the deep ocean, where it would remain safely sequestered. An additional benefit of Kruger's plan is that adding lime to the ocean could also counteract the ocean acidification we're seeing today. Given that geologists have ample evidence that previous mass extinctions were associated with ocean acidification, geoengineering an ocean with lower acid levels is obviously beneficial. "A caveat is that we don't know what the environmental side effects of this would be," Kruger said, echoing the refrain I'd already heard from Driscoll and others.

Kruger's idea depends on something that the algae plan does as well. It's called ocean subduction, and it refers to the slow movement of chemicals between the upper and lower layers of the ocean. Near the ocean surface, oxygen and atmospheric particles are constantly mixing with the water. When this layer becomes saturated with carbon, we see carbon levels rise in the atmosphere because the ocean can no longer act as a carbon sink. But the lower reaches of the ocean can sequester massive amounts of carbon beyond the reach of our atmosphere. "If the ocean were well mixed overall we wouldn't have the problem with climate change," he said. "But the interaction between the deep ocean and the surface is on a very slow cycle." The goal for a lot of geoengineers is to figure out how to sink atmospheric carbon deep down into the water, where a lot of it will eventually become sediment. Kruger's limestone plan wouldn't deliver the carbon directly to the depths, the way the algae plan might have. Instead, the lime would keep more carbon locked into the upper layers of the ocean, allowing time for the ocean's subduction cycle to carry more of it down into the deep.

Another possible method of pulling carbon down with rocks is called "enhanced weathering." In chapter two, we saw how intense weathering from wind and rain on the planet during the Ordovician period actually wore the Appalachian Mountains down to a flat plain. Runoff from the shrinking mountains took tons of carbon out of the air, raising oxygen levels and sending the planet from greenhouse to deadly icehouse. The Cambridge physicist David MacKay recommends this form of geoengineering in his book *Sustainable Energy—Without the Hot Air*. "Here is an interesting idea: pulverize rocks that are capable of absorbing CO_2, and leave them in the open air," he writes. "This idea can be pitched as the acceleration of a natural geological process." Essentially, we'd be reenacting the erosion of the Ordovician Appalachians. MacKay imagines finding a mine full of magnesium silicate, a white, frangible mineral often used in talcum powder. We'd spread magnesium silicate dust across a large area of landscape or perhaps over the ocean. Then the magnesium silicate would quickly absorb carbon dioxide, converting it to carbonates that would sink deep into the ocean as sediment.

However we do it, enhanced weathering relies on the idea that we could take advantage of the planet's natural geological processes to maintain the climate at a temperature that's ideal for human survival. Instead of allowing the planet's carbon cycle to control us, we would control it.

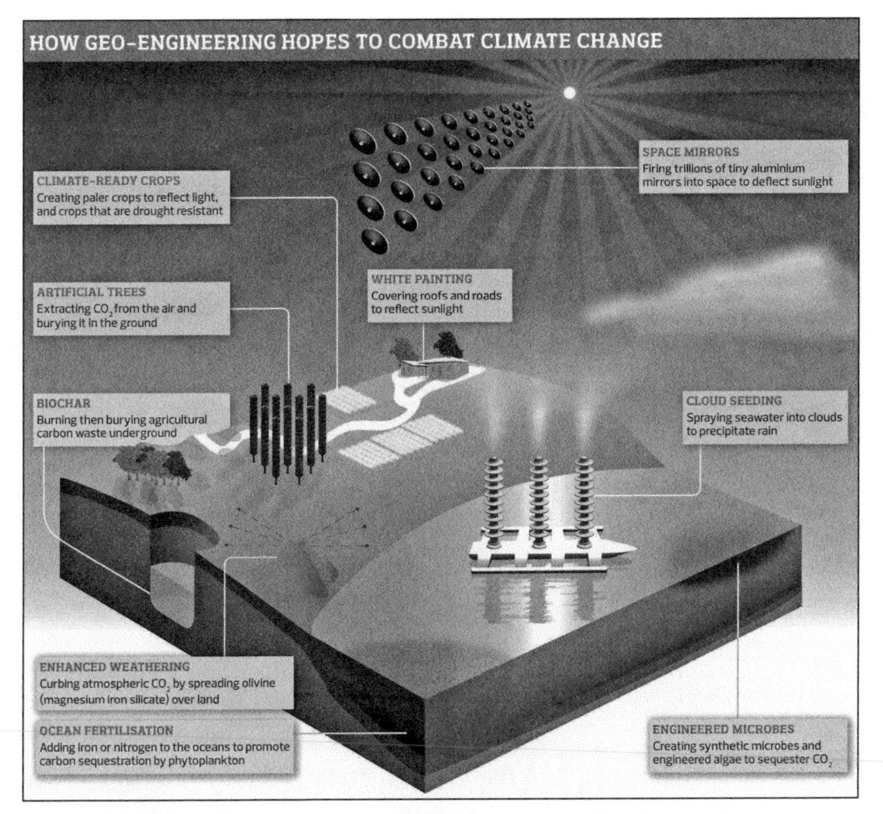

This diagram illustrates how a number of geoengineering projects could be used to transform the Earth's climate by drawing carbon out of the atmosphere and reflecting sunlight back into space.

We would adapt the planet to our needs by using methods learned from the Earth's history of extraordinary climate changes and geological transformations. Of course, this all depends on whether we can actually make geoengineering work.

The Moral Hazard

There is what Kruger and his colleagues call a "moral hazard" in doing geoengineering research, because it could popularize the idea that geoengineering solutions are a magical fix for our climate troubles. If policymakers believe that there's a "cure" for climate change just around the corner, they may not try to cut emissions and invest in sustainable energy. "It's as if a scientist had some good results while testing a cancer cure in mice, and we started telling kids, 'Hey, it's OK to smoke, we're about to cure cancer,' " Kruger said. The point is that we're very far from being certain that geoengineering would work, and until we've got a lot more hard data, we have to assume that the best way to slow down climate change is to stop using fossil fuels.

There's another worry, too. "There's a potential for nation-states to see geoengineering activities as a threat," Cascio cautioned. Harking back to what Driscoll said about how stratospheric reflective particles might cause cooling in some places, but warming in others, Cascio warned that solar management might cause famines in some regions of the world while others cool down into fruitful growing seasons. So one country's climate solution might be another one's downfall. A failed experiment in stratospheric particle injection might not just be horrible weather—it might be nuclear retaliation from countries who feel attacked.

To deal with these moral and political hazards, Kruger and several colleagues created the Oxford Principles, a set of simple guidelines for geoengineers to follow in the years ahead. Spurred by a call from the U.K. House of Commons Science and Technology Committee, Kruger met with a team of anthropologists, ethicists, legal experts, and scientists to draft what he called "general principles in the conduct of geoengineering research." The Oxford Principles call for:

1. Geoengineering to be regulated as a public good.
2. Public participation in geoengineering decision-making.
3. Disclosure of geoengineering research and open publication of results.
4. Independent assessment of impacts.
5. Governance before deployment.

Kruger emphasized that the principles must be simple for now, because geoengineering is still developing. First and foremost, he and his colleagues want to prevent any one country or company from controlling geoengineering technologies that should be used for the global public good. Principles 2 and 3 touch on the importance of openness in any geoengineering project. (The rogue geoengineer in Canada notably violated principle 2, getting absolutely no input from the public before seeding the waters.) Kruger feels strongly that as long as the public is informed and able to participate, they won't fear geoengineering in the way many people have come to fear other scientific projects, like GMO crops. Finally, the principles aim to prevent unchecked experimentation that could lead to environmental catastrophe, while also avoiding regulations so restrictive that they stifle innovation. Principle 5, "governance before deployment," speaks in part to Cascio's concern about countries interpreting geoengineering as an attack. Before we turn the skies white, or fill the oceans with lime, we must form a governing body that allows nations and their publics to consent to change the fundamental geological processes of the world. "There are huge risks associated with doing this, but doing nothing has huge risks as well," Kruger concluded.

To make Earth habitable for another million years, we will have to start taking responsibility for our climate in the same way we now take responsibility for hundreds of thousands of acres of farmland. Geoengineering of some kind is critical for our survival, because it's inevitable that our climate will change over time. Certainly we'll have to adapt to new climates, but we'll also want to adapt the climate to serve us and the creatures who share the world's ecosystems with us. If we want our species to be around for another million years, we have no choice. We must take control of the Earth. We must do it in the most responsible and cautious way possible, but we cannot shy away from the task if we are to survive.

Of course, we can't stop at the edges of our atmosphere. If climate change doesn't extinguish us, an incoming asteroid or comet could. That's why we're going to have to control the volume of space around our planet, too. We'll find out how that would work in the next chapter.

20. NOT IN OUR PLANETARY BACKYARD

WE ALREADY KNOW what an asteroid strike did to the creatures who lived during the Cretaceous period. Though it may not have been the sole driver of the K-T mass extinction, the 6.2-mile-diameter bolide that landed off the coast of Mexico roughly 65 million years ago devastated the planet, radically altering the Earth's climate for possibly a decade or more. Among the scientists who study impacts, that one would have been classed as a 10 out of 10 on the Torino scale, a kind of Richter scale used to quantify impact hazards. Such disasters, where the entire planet is affected, are likely to strike once every 100,000 years or so (though not necessarily with the destructiveness of the K-T impact). That means we are long overdue for another one.

Will we wake up tomorrow to a newscaster telling us that humanity has six months to live, so we'd better make the best of things before an asteroid wipes us out?

Not likely. Contrary to Hollywood myths, we'd probably see an asteroid like the one that hit during the K-T mass extinction coming many years before it smashed into us. Less than two decades after scientists discovered the role an asteroid played in the planet's most recent mass extinction, NASA launched an asteroid-spotting program called Spaceguard. The goal of Spaceguard was to discover and track 90 percent of near-Earth objects larger than a kilometer. A near-Earth object, or NEO, refers to asteroids, meteoroids, comets, and other heavenly bodies whose orbits around the sun bring them close to our own orbit. Most NEOs are not dangerous—they're either so small that our atmosphere would burn them up, or they zip past us millions of kilometers away. That being said, there is a class of NEO called potentially hazardous objects, or PHOs, and these are the ones we have to be worried about. To achieve PHO status, an object has to be larger than 1 km and its likely trajectory must take it closer than 7,402,982.4 km from Earth.

That sounds pretty far away, especially when you consider that we've had some near misses over the past two decades when sizable asteroids have come within thousands of kilometers of the planet (some would have caused explosions comparable to a nuclear bomb if they had hit). But our solar system is a constantly shifting set of gravitational fields, and the orbits of small objects shift a lot over time. If an asteroid zooms past Jupiter or another planet on its way to us, gravity from that other body could easily pull the asteroid into a new course, converting it from distant to deadly. That's why astronomers want to keep a sharp eye on any large rocks or balls of ice that come within a few million kilometers of our orbit.

The good news is that over the past two decades, we've gotten pretty good at spotting and tracking NEOs and PHOs. The bad news is that, at least right now, nobody is quite sure what we'd do in what NASA astronomer and asteroid hunter Amy Mainzer calls one of the most hopeful scenarios. That would be when an astronomer—possibly Mainzer herself—verifies tomorrow that there's a mile-diameter asteroid twenty years out, on a direct collision course with Earth.

Preparing for an Asteroid Hit

Mainzer is obsessed with seeing into space. That's why she's worked on instrumentation for NASA spacecraft like the WISE (Wide-field Infrared Survey Explorer) satellite, whose sole job was to map as much of the sky as possible using an infrared telescope. Once the WISE mission was complete, Mainzer and her colleagues were able to reprogram the craft in 2010 to scan the sky for NEOs—they dubbed this mission NEOWISE. It was the NEOWISE mission that helped complete the Spaceguard project by identifying enough one-kilometer-or-bigger NEOs that we can say with confidence that we now know where 90 percent of them are. In all, we've located nearly 900 NEOs of that size. "That's good for Earthlings," Mainzer told me lightly by phone from her office at the Jet Propulsion Lab in California. But then, more seriously, she added, "We don't know where most of the other ones are." In her most recent work, Mainzer gathered data on PHOs among asteroids, and she and her colleagues estimate there might be as many as 4,700 of these potential impactors that are bigger than 100 meters. To give you a sense of what that means, a 100-meter asteroid wouldn't cause a mass extinction, but it would easily flatten a city or a small country. If it landed in the ocean, the tsunamis it generated could do profound damage to coastal areas.

Given that our local volume of space is swarming with deadly rocks, why aren't we bombarded all the time? The simple answer is that we are. Every day, we are hit by tiny NEOs, most of which we never notice because they flame out before reaching the Earth's surface. "You know the video game Asteroids?" Mainzer asks. Of course I do. "Well, it's actually pretty accurate. Asteroids break up and make more little pieces. And there are far more little pieces than big pieces." Aside from the relative rarity of larger asteroids, there's also the fact that our solar system is a dynamic, constantly shifting sea of debris. All the overlapping gravitational fields of the planets and their moons may send rocks spinning into our path, but they also send them spinning out of it, too. "If you put a particle in near-Earth space, it doesn't stay stable," Mainzer explained. "After about ten million years, it will go into the outer solar system, crash into the sun, or crash into the Earth." Keep in mind that 10 million years is nothing to a planet like Earth, which has been around for 4.5 billion years. Essentially, there's only a short time window for these NEOs to do any damage before they're hurled elsewhere by gravity.

Still, Mainzer notes, there are probably "source regions" of the asteroid belt that are constantly resupplying the inner system with new NEOs.

Possibly these source regions are shooting out new NEOs because of gravitational resonances with Mars and Jupiter, the two planets whose orbits sandwich the asteroid belt. "I like to think of it as a flipper on a pinball machine," Mainzer said. "That's what these resonances are like in the main belt—if an asteroid drops into one, it can get hurled very far from its original location."

With the amount of data we've gathered from satellites like NEOWISE, it's reasonable to hope we'd have twenty years to deal with an asteroid or other PHO big enough to cause destruction over the entire Earth. Knowing where most of the large NEOs are can help astronomers to track their movements and determine whether they're on a collision course. That said, collision courses are always expressed in probabilities. We can't predict precisely where gravity will tug one of these objects on its way to our cosmic neighborhood. Also, we're still struggling to track objects that could cause tremendous damage without actually destroying humanity. "Your warning time depends on the design of your instruments," Mainzer said. She's currently working on plans for a new space telescope, dubbed NEOCam, designed to spot objects smaller than 100 meters and to find more of those PHOs. "We're designing it to give us decades of warning," she said. The goal for Mainzer and others in her field is to get 20 to 30 years of warning for a likely impact, so that we have as many options as possible for stopping it.

Defending the Planet

Most people who are serious about defending Earth from PHOs don't talk about blowing things up. As Mainzer explained with the 8-bit game Asteroids, the problem is that asteroids tend to break down into smaller asteroids. Nuking an incoming object might not do much more than shower our planet with dozens of burning chunks rather than one big one. The damage would be roughly the same. The reason Mainzer's data-gathering is so crucial is that the further away an asteroid is when we spot it, the easier it will be to nudge it out of the way. That's right—our best bet is to nudge it. "Blowing up asteroids may be fun, but an Aikido move would be better," Mainzer said, only half joking. "Having time gives you the ability to move its trajectory without a lot of energy."

The question of how to finesse this Aikido move in space has been the longtime concern of a loose coalition of scientists, policy-makers, and government representatives associated with the Center for Orbital and Reentry Debris Studies. Run by aerospace engineer William Ailor, the Center has developed a series of suggestions over the past 15 years for how we'd deal with asteroid threats. An affable man with tidy gray hair and a touch of the South in his speech, Ailor sketched out how he thought an impact scenario might unfold. "Anyone can find these things," he said. "There are amateur astronomers all over, as well as more formal programs in space agencies. Most likely, it would be spotted by that community." If it's a smaller object, we might have very little time to prepare. "People like to think we'll have twenty years, but we might only have a few years."

The next big hurdle wouldn't be the question of how to divert the asteroid, though. Say, for example, Mainzer's NEOCam is in orbit and her team spots an object bigger than 1 km that has a 1-in-50 probability of smashing into the Earth. "Should we spend money on that now?" Ailor asked. "Given the fact that it takes you years to build a new payload and fly a mission out to do something, you may have to start spending money before you're certain it's going to hit. And that's the challenge for decision-makers." The problem is that every PHO is a probability... until it isn't. And the time to act decisively to push an incoming object out of the way is almost inevitably going to be long before we can establish that a collision is a certainty. Meanwhile, as the object hurtles closer to us in space, the less likely it is that we'll be able to gently nudge it into a new orbit, out of our way.

So who would have to step up and push the world to launch anti-PHO spacecraft? The U.N. Committee on the Peaceful Uses of Outer Space has a group called Action Team-14 that deals with NEOs, and would likely be the first agency to coordinate Earth's defense in this situation. Provided they can get buy-in from countries and corporations with the means to build spacecraft for the mission—and that's a big if—the group would have to decide exactly what method of PHO deterrence would work best.

Ailor's company, the Aerospace Corporation, did a study in 2004 on what would be required to take a 200-meter object from a 1-in-100 probability of hitting Earth to 1 in 1,000,000. "You have to launch quite a few spacecraft," Ailor said. "There's a misconception that you would send up just one vehicle." Redundancy would be crucial, in case one of the crafts fails—and besides, some techniques for moving the object require multiple spacecraft to work. Also, despite what we saw in the asteroid-nuking flick *Armageddon,* the vehicle would be a remote-controlled robotic craft. "If a human can get there, it's way too close," Ailor asserted.

If we have enough time, we'd want to try what Ailor called slow-push techniques. One would involve using a swarm of small spacecraft equipped with lasers designed to boil material off the surface of the object. As the PHO spat pieces of itself into space, enough thrust would be generated to gently move it out of its deadly path toward Earth. Another possibility would be to create a "gravity tractor" with one or more spacecraft. Parking bulky objects like other asteroids or big spacecraft near the distant object

In this image by Ron Miller, we see a probe pushing an NEO out of Earth's path.

might generate enough gravitational pull to move it just enough. Many years later, this small perturbation would elegantly divert its course into a completely harmless orbit. Both of these techniques are untested. But as more spacecraft venture to NEOs and the asteroid belt over the next decade, we're likely to see experiments to test whether these techniques could, in fact, jar a large object out of its current orbit.

What if the asteroid were heading toward us today, and we hadn't had a chance to test the slow-push systems? "We don't have anything off the shelf other than a kinetic impactor," Ailor said casually, as if he were talking about computer parts. A kinetic impactor is "basically hitting it with a rock," he explained. We've already tried this method on a comet with NASA's Deep Impact mission, when a probe hit the Tempel 1 comet with a giant copper slug, dislodging huge amounts of dust and ice. Tempel 1's orbit was perturbed slightly. So we know for certain that if we hit an incoming object with slugs or rocks, we have a good chance of redirecting it. "If you have one that gets too close or is bigger, you might have to use a nuke to move it," Ailor conceded. That's a last resort, and also untested.

The problem is that even our "off the shelf" kinetic-impactor solution would be tough. "You'd have to pull a craft together, grab the right kind of payload to do what you wanted, and find a launch pad," Ailor said, seeming to be mentally ticking off a list he'd pondered many times. On top of that, there would be the issue of how to inform the public without causing either mass panic or denial. It's easy to imagine people voting against an expensive anti-PHO program if there were only a 1-in-500 chance of mass extinction. Still, it's possible we might band together as a civilization to deal with this existential threat, and fail anyway. As Ailor put it, "Of course, you might miss."

In the Event of Fallen Civilization, Please Open This File

If we're facing an impact that's a 10 on the Torino scale—that is, from an asteroid comparable to the one that hit at the K-T—we are certainly facing a mass extinction. The world would be wrapped in fires, and cities would be shaken by quakes, broiled by volcanic eruptions, and flooded by tsunami waters. Over the long term, the climate would be transformed by aerosols thrown into the stratosphere. How would we survive?

Initially, our survival would depend on retreating to the kinds of underground cities we discussed in chapter 17. The immediate aftereffects of the hit would be similar to a massive nuclear war, minus the radioactive fallout. Underground, we would be relatively safe from the worst of the firestorms and other disasters. Aboveground, temperatures and fires would die down relatively quickly. Within weeks, we'd be able to poke our heads back up and see the roiling clouds of dust that had replaced our sky. And that's when our real troubles would begin. We'd likely suffer through something like a nuclear winter. Alan Robock, the atmospheric scientist who warned against solar-management geoengineering with particles in the stratosphere, was among the first scientists to suggest that supermassive explosions would result in planetary cooling. And the cold would likely intensify for several years. In an early paper about nuclear winter, Robock outlines a scenario that sounds like a mild icehouse. The first year after the explosion–in this case, an asteroid strike–we'd see a global buildup of ice and snow and lowering of temperatures by about two degrees. But as the cold deepened, the planet's snowy surface would reflect even more light–creating a runaway effect that would cool us down possibly as much as 15 or 16 degrees in the following several years.

Without sunlight, agriculture would grind to a halt and wild plants would die back. Herbivores would die, and then the carnivores who fed on them would die out, too. Creatures who dwelled near the surface of the water would suffer in the immediate effects of the hit. Then, over time, runoff from the decimated land would fill the oceans with carbon and create deadly pockets of anoxic waters. Humans would have to rely on greenhouses for food, as well as whatever we could cultivate with little sunlight. Mushrooms, fungus, and insects would play a much bigger role in our diets than they do today.

There is also the distinct possibility that enough people would be killed in the strike that it would be impossible to maintain our civilization at its current level of development and energy needs. Megacities and high-tech societies require many people with specialized knowledge to make them function, and if only a few million people are left alive on the planet, it's unlikely that we'll have the right combination of skills to resurrect New York or Tokyo. What would we do if we had to rebuild human civilization from scratch? This is the kind of question that dogs apocalyptic science fiction, but preoccupies people in the real world, too. Alex Weir, a software developer based in Zimbabwe, is part of a small group that maintains the CD3WD database, a relatively small set of computer files that contain as much human knowledge as possible about what amounts to a pre-technological civilization. There are sections devoted to basic medicine, agriculture, town building, and power generation. At 13 gigabytes, it's easily stored on a few DVDs, or (ideally) printed out as a thick sheaf of papers and stored in a three-ring binder. The idea is to keep the CD3WD database in your survival kit, a backup copy of everything history has taught us about creating an early industrial society. It is one of the simplest and most profound examples of how survival requires us to remember what has come before. If people need guidance with rebuilding the world after the icehouse is over, CD3WD and similar projects can help us restart civilization as quickly as possible.

It is inevitable that the Earth will be on a collision course with a PHO at some point. Obviously, our first duty is to keep mapping the skies, tracking NEOs, and perfecting our asteroid-nudging technologies. But we also need to accept that the Earth isn't the safest place for us if we want to survive for another million years. We need to scatter to other planets and moons, building structures in space so that even if Earth is wiped out, humanity will survive. That's why one of the keys to long-term existence involves creating devices that will help us escape the planet. One such device is the subject of the next chapter.

21. TAKE A RIDE ON THE SPACE ELEVATOR

EVENTUALLY WE'LL HAVE to move beyond patrolling our planetary backyard and start laying the foundations for a true interplanetary civilization. Asteroid defense and geoengineering will only take us so far. We need to scatter to outposts and cities on new worlds so that we're not entirely dependent on Earth for our survival–especially when life here is so precarious. Just one impact of 10 on the Torino scale could destroy every human habitat here on our home planet. As horrific as that sounds, we can survive it as a species if we have thriving cities on Mars, in space habitats, and elsewhere when the Big One hits. Just as Jewish communities managed to ensure their legacy by fleeing to new homes when they were in danger, so, too, can all of humanity.

The problem is that we can't just put our belongings into a cart and hightail it out of Rome, like my ancestors did when things got ugly in the first century CE. Currently, we don't have a way for people to escape the gravity well of planet Earth on a regular basis. The only way to get to space right now is in a rocket, which takes an enormous amount of energy and money–especially if you want to send anything bigger than a mobile phone into orbit. Rockets are useless for the kind of off-world commuter solution we'll need if we're going to become an interplanetary civilization, let alone an interstellar one. That's why an international team of scientists and investors is working on building a 100-kilometer-high space elevator that would use very little energy to pull travelers out of the gravity well and up to a spaceship dock. It sounds completely preposterous. How would such an elevator work?

That was the subject of a three-day conference I attended at Microsoft's Redmond campus in the late summer of 2011, where scientists and enthusiasts gathered in a tree-shaded cluster of buildings to talk about plans to undertake one of humanity's greatest engineering projects. Some say the project could get started within a decade, and NASA has offered prizes of up to $2 million to people who can come up with materials to make it happen.

The physicist and inventor Bryan Laubscher kicked off the conference by giving us a broad overview of the project, and where we are with current science. The working design that the group hopes to realize comes from a concept invented by a scientist named Bradley Edwards, who wrote a book about the feasibility of space elevators in the 1990s called *The Space Elevator.* His design calls for three basic components: A robotic "climber" or elevator car; a ground-based laser-beam power source for the climber; and an elevator cable, the "ribbon," made of ultra-light, ultra-strong carbon nanotubes. Edwards's design was inspired, in part, by Arthur C. Clarke's description of a space elevator in his novel *The Fountains of Paradise.* When you're trying to take engineering in a radical new direction that's never been tried before, sometimes science fiction is your only guide.

What Is a Space Elevator?

A space elevator is a fairly simple concept, first conceived in the late nineteenth century by the Russian scientist Konstantin Tsiolkovsky. At that time, Tsiolkovsky imagined the elevator would look much like the Eiffel Tower, but stretching over 35,000 kilometers into space. At its top would be a "celestial castle" serving as a counterweight.

A century after Tsiolkovsky's work, Bradley speculated that a space elevator would be made of an ultra-strong metal ribbon that stretched from a mobile base in the ocean at the equator to an "anchor" in geostationary orbit thousands of kilometers above the Earth. Robotic climbers would rush up the ribbons, pulling cars full of their cargo, human or otherwise. Like Tsiolkovsky's celestial castle, the elevator's anchor would be a counterweight and space station where people would stay as they waited for the next ship out. To show me what this contraption would look like from space, an enthusiast at the Space Elevator Conference attached a large Styrofoam ball to a smaller one with a string. Then he stuck the larger ball on a pencil. When I rolled the pencil between my hands, the "Earth" spun and the "counterweight" rotated around it, pulling the string taut between both balls. Essentially, the rotation of the Earth would keep the counterweight spinning outward, straining against the elevator's tether, maintaining the whole structure's shape.

Once this incredible structure was in place, the elevator would pull cargo out of our gravity well, rather than pushing it using combustion. This setup would save energy and be more sustainable than using rocket fuel. Getting rid of our dependence on rocket fuel will reduce carbon emissions from rocket flights, which today bring everything from satellites to astronauts into orbit. We'll also see a reduction in water pollution from perchlorates, a substance used in making solid rocket fuel, and which the Environmental Protection Agency in the United States has identified as a dangerous toxin in our water supplies.

A space elevator would be a permanent road into space, making it possible for people to make one or more trips per day into orbit. Passengers could bring materials up with them so that we could start building ships and habitats in space. Once we started mining and manufacturing in space, elevators would be used to bring payloads back down, too. Most important, a working space elevator is many thousands of times cheaper than the one-time-use Soyuz rockets that bring supplies to the International Space Station, only to destroy themselves in Earth's atmosphere. NASA reports that each Space Shuttle launch cost about $450 million. Much of that money was spent on storing enough fuel to complete the round-trip back to Earth. But groups working on space-elevator plans believe their system could reduce the cost of transporting a pound of cargo into space from today's $10,000 price tag to as little as $100 per pound.

In this illustration by Pat Rawlings for NASA, you can see the climber in the foreground and the tether stretching back down toward distant Earth.

Getting Ready to Build

The elevator would be attached to the Earth at the equator, where geostationary orbit happens, probably on a floating platform off the coast of Ecuador in international waters. This is a likely building site because it is currently an area of ocean that experiences very little rough weather, and therefore the elevator could climb out of our atmosphere with as little turbulence as possible. According to Edwards's plan, the elevator ribbon would stretch 100,000 kilometers out into space (about a quarter of the distance to the Moon), held taut by a counterweight that could be anything from a captured asteroid to a space station. A ride up would take several days, and along the ribbon would be way stations where people could get off and transfer to orbiting space stations or to vessels that would carry them to the Moon and beyond.

The elevator car itself is the easiest thing for us to build today. It would be an enormous container, with atmospheric controls for human cargo, connected to large robotic arms that would pull the car up the ribbon hand over hand. We already have robotic arms that can scale ropes and lift incredibly heavy objects. This aspect of the space elevator is so widely understood that the Space Elevator Conference sponsored a "kids' day" that included LEGO space-elevator-climber races. Robots designed by teens and kids competed to see which could climb "ribbons" attached to the ceiling and place a "satellite" at the top.

Of course it will take some effort to get from LEGO climbers to lifters big enough to haul components of a space hotel up through thousands of kilometers of atmosphere and space. But this is within the capabilities of our current industrial technology. So we've got our elevator car. But how will it be powered?

One of the many arguments in favor of the elevator concept is that it will be environmentally sustainable. The dominant theory among would-be space-elevator engineers at this point is that we'll install lasers on the space-elevator platform, aimed at a dish on the elevator that will capture the beam and convert it to power. This technology is also within our reach. In 2009, NASA awarded $900,000 to LaserMotive for its successful demonstration of this so-called "wireless power transmission" for space elevators. In 2012, NASA offered a similar prize for a power-beaming lunar rover. The biggest problem with the power-beaming idea currently is that we are still looking at fairly low-power lasers, and as the space elevator ascended higher into the atmosphere the beam from such a laser would scatter and be blocked by clouds. It's possible that only 30 percent of the beam would reach the dish once the elevator was in space.

Still, we have seen successful demonstrations of power beaming, and companies are working on refining the technology. We don't quite have our perfect power beam yet, but it's on the way.

The Missing Piece: An Elevator Cable

At the Space Elevator Conference, participants devoted an entire day to technical discussions about how we'd build the most important part of the space elevator: its cable, often called the ribbon. Again, most theories about the ribbon come from Edwards's plans for NASA in the 1990s. At that time, scientists were just beginning to experiment with new materials manufactured at the nanoscale, and one of the most promising of these materials was the carbon nanotube. Carbon nanotubes are tiny tubes made of carbon atoms that "grow" spontaneously under the right conditions in specialized chambers full of gas and chemical primers. These tubes, which look a lot like fluffy black cotton, can be woven together into ropes and textiles. One reason scientists believe this experimental material might make a good elevator cable is that carbon nanotubes are theoretically very strong, and can also sustain quite a bit of damage before ripping apart. Unfortunately, we haven't yet reached the point where we can convert these nanoscopic tubes into a strong material.

Carbon nanotube material is so light and strong that the elevator cable itself would be thinner than paper. It would literally be a ribbon, possibly several meters across, that the robotic cars would grip all the way up into space. Every year at the Space Elevator Conference, people bring carbon nanotube fibers and compete to see which can withstand the greatest strain before breaking. Winners stand to gain over a million dollars from NASA in its Strong Tether Challenge. Sadly, the year I attended, nobody had fibers that were strong enough to place (but there's always next year!).

Researchers from the University of Cincinnati and Rice University, where there are nanomaterials labs investigating the tensile strength of carbon nanotubes, explained that we are years away from having a working elevator ribbon made of carbon nanotubes. Though the microscopic tubes on their own are the strongest material we've ever discovered, we need to make them into a "macromaterial"—something that's big enough to actually build with. And making that transition into a macromaterial can be difficult, as the University of Cincinnati chemical engineer Mark Haase explained:

> I like to compare [carbon nanotube development] to the development of aluminum in the first half of the twentieth century. In the years prior to this, aluminum had been known, and it was available in small labs. It was rare and expensive, but there was interest in it because it had strange properties. It was very valuable because of this. As the twentieth century started to progress, we developed the infrastructure and the technology as well as an understanding of the material itself that allowed us to mass-produce aluminum. And that's when we started to see it infiltrating modern life in airplanes, consumer goods, and more. Carbon nanotubes are at that early stage—it's an interesting material but very difficult and expensive to make. However, I and some of my colleagues are working on making those breakthroughs so that, much like aluminum in the second half of the twentieth century, we can develop a material that will change the modern landscape.

Haase added that the barrier here is that we need to invent an entirely new material, and then figure out how to string it between the Earth and a counterweight without it breaking. That's not a trivial problem, even once we reach the point where we can create a carbon nanotube ribbon. What if a huge storm hits while the elevator is climbing into the stratosphere? Or what if one of the millions of pieces of junk orbiting the Earth, from bits of wrecked satellites to cast-off chunks of rockets, slams into the elevator ribbon and rips it? This may be an enormous structure, but it will have some vulnerabilities and we need to determine how we'll protect it.

How do you dodge an incoming piece of space junk that's headed right to your elevator ribbon? Engineer Keith Lofstrom suggested mounting the ribbon on a massive maglev platform designed to move the line in any direction very rapidly, basically yanking it out of the way. Rice University materials-science researcher Vasilii Artyukhov argued that we might not want to use carbon nanotubes at all, because they break in several predictable ways, especially when they're under constant strain and bombarded with cosmic rays from the sun. He thought an alternative material might be boron nitride nanotubes, though these are even more experimental than carbon nanotubes at this point.

Ultimately, the elevator cable is our stumbling block in terms of engineering. But there are also social and political issues we'll have to confront as we begin our journey into space.

Kick-starting the Space Economy

Building the elevator goes beyond engineering challenges. First, there's the legal status of this structure. Who would it belong to? Would it be a kind of Panama Canal to space, where everybody pays a toll to the country who builds it first? Or would it be supervised by the U.N. space committees? Perhaps more urgently, there is the question of how any corporation or government could justify spending the money to build the elevator in the first place.

One of the world experts on funding space missions is Randii Wessen, an engineer and deputy manager of the Project Formulation Office at the Jet Propulsion Laboratory. An energetic man with a quick wit, Wessen has a lifetime of experience working on NASA planetary exploration missions, and now one of his great passions is speculating about economic models that would support space flight. We've recently witnessed the success of Elon Musk's private company SpaceX, whose Falcon rocket now docks with the International Space Station, essentially taking on the role once played by the U.S. government–funded Space Shuttles. "The bottom line is that you need to find a business rationale for doing it," Wessen told me. "What I would do is parallel the model that was used for the airplane." He swiftly fills in a possible future for commercial spaceflight, by recalling how airplanes got their start:

> The first thing that happens is the military wants one—they'll fund it themselves. Next the U.S. government says this is critical to national security or economic competitiveness, so we need to make up a job for these guys to keep them in business. For airplanes, the government said, "We'll have you deliver mail." They

> didn't need this service, but they gave it to airline companies to keep them going. This is analogous to spacecraft today. The government is saying [to companies like SpaceX], "We want you to resupply the space station." That's where we are now. As this gets more routine, these private companies are going to say, "If we put seats on this thing, we'll make a killing." They did it with airplanes. You can see that starting today, with four or five different companies who have suborbital and orbital launch capability.

Like many other people in the slowly maturing field of commercial spaceflight, Wessen is convinced that government contracts and tourism represent the first phase of an era when sending people to space is economically feasible. He noted that SpaceX's founder, Musk, has said it's reasonable to expect payload costs to go down to roughly $1,000 per kilogram. "Everything cracks open at that point," Wessen declared. SpaceX isn't the only private company fueling Wessen's optimism. Robert Bigelow, who owns the Budget Suites hotel chain, has founded Bigelow Aerospace to design and deploy space hotels. In the mid-2000s, Bigelow successfully launched two test craft into orbit, and he is now working on more permanent orbiting habitats. Meanwhile, Moon Express, a company in Silicon Valley, is working closely with NASA and the U.S. government to create crafts that could go to the Moon. Its founders hope to have a working prototype before 2015.

Google is another Silicon Valley mainstay that is investing in the burgeoning space economy. The company recently announced its Google Lunar X Prize, which will award up to $30 million to a privately funded company that successfully lands a robot on the Moon. To win the prize, the robot must go at least 500 meters on the Moon's soil, called regolith, while sending video and data back to Earth. Alex Hall, the senior director of the Google Lunar X Prize, described herself as "the Lunar Chamber of Commerce." At SETICon, a Silicon Valley conference devoted to space travel, Hall told those of us in the audience that the Lunar X Prize is "trying to kick-start the Lunar Space Economy." She said the group measures its success not just in robots that land on the Moon, but in creating incentives for entrepreneurs to set up space-travel companies in countries where no orbital launch facilities have existed before. Mining and energy companies are among the groups most interested in what comes out of the Google X Prize, she said. The X Prize "is the first step to buying a ticket to the Moon, and using the resources on the Moon as well as living there." Bob Richards, a cofounder of Moon Express, is one of the contenders for the Google X Prize. He spoke on the same panel as Hall at SETICon, and amplified her arguments. "This isn't about winning—it's about creating a new industry," he explained. "We believe in a long-term vision of opening up the Moon's resources for the benefit of humanity, and we're going to do it based on commercial principles."

The space elevator is the next stage in the space economy. Once we have a relatively cheap way of getting into orbit, and a thriving commercial space industry partly located on the Moon, there will be a financial incentive to build a space elevator—or more than one. It may begin with funding from governments, or with a space-obsessed entrepreneur who decides to invest an enormous amount of money in a "long-term vision" of the kind Richards described. Already, we see the first stirrings of how such an arrangement might work, with a future Google or Budget Suites providing the initial capital required to move the counterweight into place, drop the ribbon from space down to the ocean, and get the beam-powered robotic climber going.

Once we've got a reliable and sustainable method of leaving the planet, we can begin our exodus from Earth in earnest. The space elevator, or another technology like it, could be the modern human equivalent of the well-trodden path that took humans out of Africa and into what became the Middle East, Asia, and Europe. It's the first leg on our next long journey as we scatter throughout the solar system.

22. YOUR BODY IS OPTIONAL

MOST OF US can imagine humans living in a future full of space elevators, and even cities on the Moon. But we usually picture our distant progeny in that future looking exactly the way we do now, the way people do in *Star Trek*. And yet of all the possible futuristic scenarios we've explored in this book, our continued evolution as a species is one of the most certain. We are going to evolve into creatures different from humans today—perhaps as different as we are from *Australopithecus*. The question is just how fast this will happen, and whether we'll use what we know about genetics to steer the process.

This concern is especially important in the context of how humans will become a space-faring civilization. We are adapted nicely to live inside the thin layer of gas surrounding the rock we call home, but in many ways that makes us terrible space travelers. First of all, the Earth's magnetic field protects us from the enormous amount of radiation in space, so our bodies never evolved a good defense against radiation damage. Solar radiation and high-energy particles would bombard our bodies on a regular basis in space and on places like the Moon and Mars, which have much weaker magnetic fields than we do at home. A person living off-world would have a high probability of developing cancer, infertility, or other radiation-induced problems. Another issue for people in space will be our extremely specific needs when it comes to sustenance. Because we draw our energy from foods native to Earth, it will never be a matter of humans colonizing space alone. We will have to bring a whole biosphere along with us, including plants and animals, as well as the exact mix of oxygen, nitrogen, and other gases we require to breathe. There are a host of other issues, too, such as the human body's tendency to atrophy in low gravity, and the fact that our life spans are so short that it would take several generations to reach even the closest neighboring stars.

For all these reasons and more, it's likely that the project of exploring space will involve a parallel project to adapt our bodies to life in environments radically different from the one where we evolved. It's possible that we'll become cyborgs, beings who are half biological and half machine. We may tweak our genomes to be radiation-resistant. Or we may, according to some futurists, become so technologically advanced that we'll be able to convert the galaxy into a giant version of Earth. No matter what happens, the humans who live in space will be different from the humans on Earth today.

Changing Ourselves to Live in Space

How would we go about modifying ourselves to be more space-worthy? If anyone would know, it would be a synthetic biologist. In chapter 18, we explored how synthetic biology could revolutionize cities by providing us with buildings that could heal themselves or even grow. The field has obvious applications for any project to create humans suited for life beyond Earth as well. Leaving aside for a moment the ethical issues of engineering space-ready humans, synthetic biology could eventually reach a point where we could accurately predict how modified human genomes would function—and then implant them in the next generation. Though such knowledge may be centuries away, it's very possible we might one day be guiding ourselves through a phase of evolution aimed at, for example, giving birth to children who could live on Mars unharmed.

To find out how this might work, I visited the UC Berkeley synthetic biologist Chris Anderson, a pioneer in the field whose research focuses in part on defining the most efficient and ethical methods for pursuing synthetic biology. A slender man with a sly smile, Anderson launched into an explanation of the future direction for synbio (as it is fondly known) by first seizing a piece of putty on his desk and vigorously smashing it between his hands. "I love this stuff!" he enthused. Apparently there is a reason the substance is sometimes marketed as "thinking putty," because Anderson's train of thought matched the pace at which he mashed out different shapes, which each looked like a new phylum of bacteria. Synbio researchers, he said, look at every organism in terms of its component parts. They don't want to engineer new life-forms—not exactly. Instead, they want to engineer *parts*, especially at the genetic level. "Fundamentally, what we're talking about is moving genes between organisms," Anderson said. "We want to write whole genomes eventually. But it's all based on components, and having the ability to predict how they will add together."

A synbiologist looks at life-forms the same way a mechanic looks at an engine. To the mechanic, the engine is a set of interoperable parts, and some of those parts could just as easily be used in another machine. Likewise, a biological part like a gene or a protein could easily be repurposed for use in another organism. In fact, Anderson pointed out, a synbio project would typically look at a particular part—a gene, for example—in its many variants across a thousand organisms. Let's say a synthetic biologist is studying a gene that thousands of plants use in photosynthesis. Her goal in that research would be to predict how the gene would function if it were used as a part in another organism. She would base her predictions on how the gene behaved in the thousand species where it currently exists. "This is a paradigm shift," Anderson said. "We've stopped focusing on studying naturally existing things and are instead building things one gene at a time. Basically, it's biological systems as a sum of their parts."

Anderson's work focuses on bacteria, and when I started to ask about modifying humans he wrinkled his nose. "I don't touch mammalian cells," he quipped. "They're such a mess." And this messiness is nothing compared with the moral quandaries presented by a future where we might use synbio to make Martians. The main problem, he said, is that we can't experiment on humans the way we do with bacteria. To make accurate predictions about what a gene will do in a given organism, you have to run thousands of tests—some of which reveal that the gene does the opposite of what you'd hoped. "If you are playing with enhancing human intelligence, for example, you might create somebody who is brain dead rather than smart," Anderson mused. "There would be so many accidents because this work involves a lot of fundamental uncertainty."

Given all the risks and their consequences, he added, "It's hard for me to see how you could even develop enough design theory to be able to safely build a human. It's not going to be socially acceptable to tweak a human in a way that could cause them to be born grossly broken. No one's going to go for that." But, he conceded, it might be possible in a future where we had "a radical transformation in our ability to predict things" on the genetic level. If we had absolute certainty about how a given gene would work in a human, then the risks would be minimal. However, Anderson was extremely dubious we'd ever reach this point. He emphasized several times that he couldn't believe that humans would ever be willing to modify our germ lines to change our species at a genetic level.

His sentiments were echoed by Claudia Wiese, a Lawrence Berkeley Lab geneticist who studies how humans respond at a cellular level to radiation in space. "During a long-duration manned space flight, substantial numbers of cells in a human body, [approximately] 30%, will be traversed by at least one highly ionizing particle track," she told me via e-mail. "Highly ionizing particles" are the most dangerous kind of radiation we can encounter in space—these energetic particles shoot through the body like infinitesimally small bullets, cutting through everything in their path, including tissues and DNA. The danger is that they would damage a cell's DNA but not kill the cell outright. Subsequently, the cell would replicate with mutated DNA, a situation that can lead to cancer.

Wiese and her colleagues believe some variants of DNA-repair genes may be better at dealing with radiation damage than others, but they're nowhere near being able to tweak these genes to make humans radiation-proof. "I think that we are a long way away from gene therapy," Wiese said. "At this point, the use of appropriate countermeasures"—like drugs such as antioxidants—"may be a more immediate and feasible way to mitigate the detrimental effects of space radiation." Still, her research and that of other geneticists working with NASA suggest that we may one day know which genes control DNA repair. Once we are able to predict the behaviors of these genes, future space-farers might tweak their genes to respond quickly and effectively when bombarded with highly ionizing particles beyond Earth's protective magnetic envelope.

But some synbio researchers aren't sure this is a good idea. Daisy Ginsberg, a London designer who works with synthetic biologists on the ethical implications of their work, said that we could take the modification of humanity way too far. Sipping tea in a London restaurant, Ginsberg's sunny disposition belied a deep pessimism. "I'm of the mind that we're going to fuck everything up," she said cheerfully. "We're going to poison the Earth and it's going to be unpleasant and expensive. I think we're going to become Morlocks." Ginsberg was referring to H. G. Wells's novel *The Time Machine,* in which the author predicts that humans will evolve into two species: the Morlocks, a hyper-technological, warlike group who live underground, and the Eloi, a dim-witted but peaceful group who are prey to the Morlocks. Ginsberg was basically predicting that humans would evolve to be hideous monsters who destroy the Earth and prey upon each other. But what about modifying ourselves to live beyond Earth, so that we stop destroying our home world? Ginsberg was dubious about that, too. "I think it's unethical to colonize space because we'll make a mess there as well," she said. "I'm sure we'll be modifying everything."

One of Ginsberg's most memorable design projects is called "pollution sensing lung tumor." It's a sculpture of a pair of human lungs, made entirely out of sparkling red crystals. It comes from one of Ginsberg's scenarios for a synbio future in which the environment is full of materials made out of biological components. One material might be a pollution sensor made from thin crystalline sheets of carbon monoxide–sensing bacteria. What if some of those biological components got loose, and started spreading in the air or water supplies? These bacterial sensors might enter the lungs of a smoker, proliferating there because of all the "pollution" they discover. Suddenly, instead of a bacterial infection that gives you a cough, you've got an infection that builds crystal sensors in your respiratory tract. The crystalline structure Ginsberg created, which at first appears to be a pair of lungs, actually represents a tumor that has been removed from a woman's body.

Ginsberg got the idea for this project after seeing plans for such sensors, as well as for materials like the self-healing concrete I described in chapter 18. Ultimately, her concern mirrored Anderson's. We might create synbio organisms with the best of intentions, but end up modifying humans in ways that make them sick or "grossly broken," as Anderson put it. At the same time, Ginsberg strongly advocated against "viewing nature as a fixed thing." Looking out the window at the busy street, she pondered out loud: "Maybe it's ethical to disrupt nature after all. It's just that there are so many questions that we don't know answers to. I don't know whether we can ever actually think it over enough to do it well."

Though neither Anderson nor Ginsberg could imagine humans modifying themselves successfully, science-fiction writers have no trouble imagining this at all. We already explored Octavia Butler's vision of a transformed humanity in her novels, and she's not the only one. The British author Paul McAuley has suggested in recent novels like *The Quiet War* that humans will modify themselves as they colonize the solar system. He imagines a war between the Inners, people like Ginsberg from the inner solar system who think it's morally reprehensible to modify their germ lines, and the Outers, people who have genetically modified themselves to adapt to life on the moons of Jupiter and beyond. McAuley began his career as a botanist at Oxford, and his background as a scientist informs his work. He told me how he thinks future humans might justify doing the human genetic experiments Anderson said "nobody would go for." It all comes down to necessity. The Outers, he told me,

> developed the [genetic] tweaks during or immediately after the political crisis that forced them to flee Earth's Moon for the Jupiter and Saturn systems; that is, the work was performed under conditions where lifeboat ethics applied. The same kind of desperation, albeit to a much lesser degree, that drives people suffering mortal

> illnesses to seek out experimental treatments. Do or die. This may explain the lack of social outrage, or the so-called "yuck" factor, in the Outer communities.

Ultimately, McAuley imagines that one possible reason we might start modifying ourselves will be because we have no other choice. We're likely to die in space anyway, so we might as well try something extreme.

Kim Stanley Robinson, another science-fiction author whose work deals with how humans will modify themselves to colonize space, told me that humans will get over the "yuck" factor as soon as we have decent longevity treatments. Once synbio researchers figure out a way to prolong human life, Robinson believes, people will be willing to experiment with their germ lines to create future humans who live for hundreds of years. And once we've done that, the floodgates will open. We'll see humans changing their bodies more radically. We might shrink ourselves into smaller creatures so that we consume fewer resources, or modify our DNA to repair itself after radiation damage in space.

It's tempting to say that scientists have the right answers here, and that the science-fiction writers are just engaging in rank speculation. But scientists have a duty to deal with strictly present-day scientific knowledge, not futuristic predictions. Sometimes it takes someone who isn't involved in the day-to-day responsibilities of scientific work to see where today's research is truly leading.

Getting Rid of Our Bodies

Synbio intervention into evolution may strike us as morally problematic, but it's a plausible outcome of today's research. Still, there are many other possible ways humanity might evolve. A group at the Oxford Martin School—where we visited a group of visionary geoengineers in chapter 19—thinks the future belongs to machines. The philosophers who make up the Future of Humanity Institute reside in a set of offices loosely clumped near a meeting room full of conference tables, a wall-sized whiteboard, a coffeemaker, and boxes of slightly odd Scandinavian sweets. Their goal is to explore existential threats to humanity, or extinction-causing events, including many of the dangers we've already discussed in this book. But their biggest concern is the possibility that we may be wiped out by "machine superintelligence," or artificially intelligent computers (AI) that essentially take over the world.

Nick Bostrom heads the institute, where he's written widely cited articles about everything from human genetic enhancement to existential threats and what he calls "the intelligence explosion." A Swedish ethicist with closely cropped hair and a perpetually serious expression, Bostrom welcomed me into a spartan office whose windows overlook the courtyard where it's said J. R. R. Tolkien wrote *The Hobbit.* When I asked him about humanity's future, he wanted to get one thing straight right away. If humanity survives, he believes, it's inevitable that we will pass through an "intelligence explosion" during the next century or two in which we will invent machines with greater-than-human intelligence. Other thinkers have called this event the "Singularity." These machines will either wipe us out or help us create a future so unlike the present that we can hardly imagine it. But why is it inevitable that we'll invent AI? "It's one of those technologies that it's hard to refrain from developing if we can," he said. "All the steps up to it have obvious uses. We want better search algorithms, better recommendations from Amazon, and automatic fraud detection. We also want to understand the human brain, for both scientific and medical reasons. It seems hard to imagine how that would stop short of a global cataclysmic event."

What Bostrom and his colleagues predict is that at some point humans will put together an advanced knowledge of the human brain with the "smart" algorithms that already power services like Google and predictive programs like the ones we've seen modeling epidemics and natural disasters. The result, he believes, will be a machine like a human brain with the processing power and memory of an enormous cluster of computers.

Imagine a brain that could process nearly unlimited information and use it to predict possible outcomes to problems and advance scientific knowledge. If such a thing existed, it would dramatically transform humanity—and our relationship to outer space. Bostom put it to me this way. Given that humans are the apex species on the planet due to intelligence, it seems likely that if we invented something with superhuman intelligence it would best us. The question is how, exactly, this besting would occur. Would it become our friendly intellectual older sibling, helping humans to stop pandemics, design the perfect space elevator, and gain superintelligence of our own? Or would it consider humans a bother, the way we do ants? In the latter case, it's possible the machine's "fatal indifference" to humanity would end our species forever. It might just kill us accidentally, as it went about its incomprehensibly advanced business.

Assuming we do make it through the intelligence explosion intact, however, Bostrom and his colleagues have a few ideas about what might happen to humanity. Most important to their scenarios is the idea of "uploading," or turning our brains into software on computers. Our minds could be transferred into virtual worlds, where we would have incredible adventures and expand our consciousness to include the whole of human knowledge. When our bodies died, our uploaded minds would live on—and maybe get downloaded into new bodies. We could also make many virtual copies of ourselves, which is a particularly weird idea that would make perfect sense in a world where you could upload your brain anytime you wanted. Why not save yourself as an upload at different times, the way you do with avatars in a video game, so that you could revert to an older copy if something horrible happened that you'd rather forget? Uploads would completely change our relationships to our bodies and identities, as we easily slid between virtual and biological existence.

Because Bostrom believes this future of superintelligence and uploads is inevitable, he's convinced that we won't go to space at all. We won't want to. Instead, we'll convert all of outer space into a giant computer running all our uploads in a vast virtual world. His idea hinges on the notion that nobody would want to live in reality anymore when they could upload themselves into a virtual world of plenitude and mental transcendence. So instead of exploring outer space in fantastical vessels, we'd use robots to dismantle every object in space, from planets and asteroids to suns and black holes. Then we'd convert these massive bodies' every molecule into a giant supercomputer where our uploaded brains could expand forever.

In essence, we'd use our superintelligence to convert all of outer space into a vast virtual space for our minds.

What would this look like? Bostrom said, "There's an image I have in my mind of... a growing sphere, a bubble of technological infrastructure with Earth at the center. It's growing in all directions at the speed of light." This growing sphere would be a machine that was converting all matter in the universe into what Bostrom calls "computational substrates," or computers powerful enough to run a simulation that would satisfy machine superintelligences. In a sense, it would be like paving over the universe with our computers. "Most likely, everybody would live in virtual reality, or some abstract reality," Bostrom concluded. Space would be ours, but only because we converted every piece of matter into our high-tech brain farm.

But, to continue the paving analogy for a moment, what if there was life using some of that matter in the universe? Wouldn't we be destroying it to build our virtual world? Bostrom is unperturbed by this possibility. "My guess is that our observable universe doesn't contain intelligence, so we don't need to worry about taking matter away from them." His main concern is what's going on inside that sphere of technology he imagined hurtling out of the Earth. What if it were a version of the Windows operating system, but with superintelligence? In a dark scenario, "we might all be paperclips or calculating pi to millions of decimals," he mused. But in a brighter one, we might be liberated from our bodies, evolving beyond death in a virtual world of our own devising. We'd have become beings who explore inner space rather than outer space.

One of Bostrom's colleagues at the institute, Anders Sandberg, was less certain that our future would be purely virtual. A gregarious man who loves science fiction, he talked just as eagerly about role-playing video games as he did about the medallion around his neck that contains instructions on how to freeze his head cryogenically in the event of his death. Sandberg shares Bostrom's belief in the intelligence explosion, allowing, however, that we might venture into outer space afterwards. But, he asserted, "having a biological body in space is stupid in many ways." He suggested we might become more like cyborgs, mechanical creatures controlled by uploaded human brains. This would protect us from radiation damage, the need for food, and many other tribulations of space travel and colonization. "Uploading is just a more flexible way of living," Sandberg explained. He suggested that we might solve the problem of how long it takes to travel in space by loading the crews' brains into software for the decades or centuries it would take to reach their destination. Once the ship arrived, those brains would be downloaded into whatever bodies might suit the planet where we arrived. Maybe those bodies would be part technological and part biological. Or they might be biological forms ideally suited for life on a world like Titan, with methane gas for atmosphere.

Excitedly mulling over our future bodies, Sandberg pointed out that living part time as software might ensure humanity's long-term survival in other ways. It could keep humanity safe from a pandemic, for example. "It's also about enhancing adults," he pointed out. "It's ethically less problematic than genetically engineering our children." So turning into machines, losing our bodies forever, may cause fewer moral quandaries than modifying our genes but keeping our bodies.

Evolution Stops for No One

Even if we don't genetically alter or upload ourselves, we will nevertheless evolve into a different sort of creature in the next million years. Many evolutionary biologists believe that humans are still evolving. The University of Chicago geneticist Bruce Lahn has demonstrated that some of our genes, like those controlling brain size, appear to be undergoing very rapid selection. And researchers in Finland have pored over the family histories recorded in a Finnish village church, and discovered clear patterns of natural and sexual selection emerging over a period of centuries among the locals. When I spoke to Oana Marcu, a SETI Institute biologist who researches how life first evolved, she emphasized strongly that we aren't the end results of billions of years of evolution. We are still in the middle of an evolutionary journey, with many changes ahead of us.

When we begin heading out into space, evolutionary pressures will select for humans best able to survive in our new environments. If there's one thing we know for sure about evolution, it's that a change in environment often leads to dramatic changes for species. It also leads to speciation. If humans spread out to many planets and moons, those groups may begin to diverge genetically after millennia have passed. No matter what scenario you think is most likely–synbio, uploads, or natural selection–our progeny may look nothing like us. But they will still be part of humanity, and they will carry in them our profound, seemingly unquenchable urge to explore new environments and adapt to them as best we can.

23. ON TITAN'S BEACH

ONE MILLION YEARS ago, our ancestors thought it was pretty fantastic to have fire and flake tools. One million years from today, humans could be living in lakefront communities on Saturn's moon Titan, using technologies that make our rocket fuels and supercomputers look like a *Homo erectus* tool kit. Astronomers often point to Titan for possible colonization because it has a thick atmosphere that could offer us some of the protection from radiation that Earth's does. Plus, it has weather very much like Earth's, with seasonal rains and snows. On Titan, there are beaches full of dunes, shimmering great lakes, and the occasional volcano. Except that the volcanoes erupt with ice and the lakes are made of methane. The spring showers are methane, too. In short, it's a place that would freeze and poison any human on Earth today. But what if, in a million years, we'd engineered humans to survive there? Maybe they would be fitted with lung implants that could convert local gases to a mix that would oxygenate their blood. They might be uploads running robotic exoskeletons, or biological beings built from genetic parts that allowed them to thrive in Titan's atmosphere. Or they might have terraformed the moon to suit human bodies.

If our progeny do make it that far, it will be because humanity has chosen exploration over warfare often enough that we've managed to work

In this image taken by the international space probe *Cassini,* we got our first glimpse of the ethane and methane lakes surrounded by sand dunes on the surface of Saturn's moon Titan.

together as a planet on several large projects. One of those projects would be pushing our species off this rock, scattering us through the solar system. This project is important for reasons that go beyond how great it would be to fly through Saturn's rings. It's also important as a long-term human goal because most of the steps that lead to its realization will take us down the pathway of survival rather than death.

Planning for Life Beyond Earth

"Our kids are the last generation who will see no city lights on the Moon," the NASA Jet Propulsion Lab's Randii Wessen told me when we talked about the economic feasibility of space travel. Though his prediction isn't outside the realm of possibility, we should also think pragmatically about the path to space. Wessen's colleague the atmospheric physicist Armin Kleinboehl is far more conservative in his estimates of when humans might live beyond Earth. Kleinboehl studies Martian weather via the Martian Reconnaissance Orbiter (MRO), a craft that has been photographing and analyzing the Martian surface from orbit since 2006. When we met, he showed me some MRO images of dust storms that periodically envelop all of Mars, making it even colder than it is normally. When I asked Kleinboehl what he thought the timeline might be for a Mars colony, he frowned and glanced over at a video simulation of Martian weather. "It won't be attractive for at least five hundred years," he said finally. "It's not very hospitable." When I asked about terraforming, Kleinboehl conceded that that was one way Mars might eventually become habitable. Perhaps we could cultivate plants or bacteria on Mars that photosynthesize, pop out free oxygen, and change the environment.

Though it's popular to imagine that humans will be exploring the galaxy in *Star Trek* style over the next couple of centuries, this leap may take a little longer. And that's good news. By the time we're ready to set up tourist resorts on Titan, we may have reached the point as a civilization that we won't "fuck everything up," as synbio designer Daisy Ginsberg so succinctly put it.

It's not very popular to suggest that the future could happen slowly, or that tomorrow's scientific innovations might take as much time as they have historically. Futurists like Ray Kurzweil are fond of suggesting that the pace of discovery is "accelerating," and that change will move at a blindingly rapid clip over the next century. While anything is possible, we shouldn't expect immortality, superintelligence, and faster-than-light travel in our lifetimes. Anticipating instantaneous, radical change diverts us from investing time in long-haul projects like building safer, sustainable cities and planning for food security. These are the kinds of scientific endeavors that can help us survive while we're waiting for somebody (or some*thing*) to invent upload technology. I'm not suggesting that we should slow down the pace of scientific discovery. Quite the opposite. I'm saying we should focus our scientific and technological energies on problems that are solvable in the near term, while always keeping our eyes on the long-term goal of exploring and adapting to worlds beyond our blue marble.

If we've learned anything from the survivors among our ancestors, it's that staying put and fighting change are not good tactics if we want to live. Survivors range over vast regions. If they encounter adversity in one environment, they try to escape and adapt to a new environment. Survivors prefer the bravery of exploration to the bravery of battle. But present-day humans differ dramatically from most of Earth's survivors in one crucial way. We can make plans for the future. And with the help of scientifically informed models, we can also consider how we would deal with many possible future scenarios. What if an asteroid hit? A flood? A plague? A drought? Right now, we have many excellent ways of figuring out how each of these species-ending disasters might unfold—and we have ways of preventing most of them from killing us all.

Surviving is partly a matter of implementing what we already know. But it's also about planning to deal with disasters we know for certain aren't survivable right now. Those are the kinds of disasters that require us to build cities on Mars, Titan, Europa, the Moon, asteroids, and any other uninhabited chunk of matter we can find. The more we explore, the more likely it is that our species will make it.

The Right Path

Perhaps no research plan expresses this idea better than the 100 Year Starship, a project run by the doctor and former astronaut Mae Jemison. The project, initially funded by the U.S. government, is now a nonprofit organization whose goal is to develop a starship capable of bringing humans to another star system. It's called 100 Year Starship because Jemison and her colleagues estimate that it will take roughly a century to develop the technology. Though this time period is short compared with the million-year view I was just describing, it's still longer than pretty much any other scientific project currently under way. I asked Jemison why she had set a goal beyond her lifetime and that of anybody working on the project now. It was both pragmatic and necessary, she said. The technologies needed are far beyond our current levels of development. And, she added, we need time "to create a movement here on Earth, imbuing society with the aspiration to get this accomplished." Calling the mission "optimistic," she added that she and the scientists associated with the project will be developing several technologies along the way to their goal which could be useful in themselves. Perhaps they'll create a better kind of propulsion, or new hydroponic systems for growing plants in space. "Weightlessness could be a platform," she mused, where we could stage any number of experiments before taking off for the next habitable star system. Her point also holds true for the science we'll develop as we prepare to build a space elevator and an asteroid-nudging fleet–it may help us back on Earth.

One of the scientists already studying what life might be like elsewhere in our solar system is the planetary scientist Nathalie Cabrol. She conducts missions for the SETI Institute, where she studies remote environments on Earth that are similar to environments we might find on Mars, Titan, or Europa. In the rocky peaks of the Andes mountains, where the atmosphere is thin, she and her team dive deep into lakes whose chemical compositions are rare for earthly bodies of water. There, Cabrol told me that they look for the kinds of life that could thrive on another world; they also try to figure out what would survive if Earth underwent a radical environmental change. Cabrol's team is making discoveries that are relevant for the near future and the far future. And in the process, they may discover something totally unexpected.

Cabrol explained that one of her current projects is to develop a robotic rover that could land on an unexplored world–like, say, in the oceans of Jupiter's moon Europa –and figure out what it should be studying. Because the robot would be on a totally new world, it would need to be able to get a baseline for what was normal there, and then judge what aspects of the environment were extraordinary or worth studying in greater detail. To explore Europa, in other words, we have to build a robot that can think like a scientist, taking in data and deciding which pieces of that data are salient. Work like Jemison's and Cabrol's led the celebrated science historian Richard Rhodes to speculate at the space exploration conference SETICon in 2012 that one unintended consequence of space exploration might be the emergence of artificial intelligence. So Nick Bostrom's intelligence explosion could happen on Europa rather than on Earth.

What's important is that the move into space sets humans on a journey that's survivable. And it's one that might yield many incredible discoveries along the way. Jemison, who told me she's a fan of Octavia Butler's writing, emphasized that she hopes the 100 Year Starship project will help humanity grapple with social as well as scientific issues. "What does it mean to be an interstellar civilization?" she asked. "What are the philosophical implications?" When I pushed her to answer those questions, however, Jemison did something very unexpected. She refused to speculate.

"The reason why is that if I speculate now, I can't keep a blank whiteboard in front of me," she explained. "As a person who is leading this, I don't want to say, 'When we get there it has to be this way.' It may be totally different from what we expect." By keeping her whiteboard blank, as it were, Jemison provides a model for what it means to plan for the future without foreclosing any possibilities. We can create maps and guides without locking ourselves into any particular outcome. The journey to the stars may take many forms. It may take centuries. But while we're waiting and researching and designing our starships, we can build a civilization that's sustainable back on Earth.

What Will They Remember About Us?

As we start our journey into the next million years, it's useful to ask what you hope your progeny will remember about *Homo sapiens*. What do you want it to mean when they call themselves "human"? When I think about my post–*Homo sapiens* offspring, frolicking with their robot bodies in the lakes of Titan, I hope they remember us as brave creatures who never stopped exploring. What unites humans of the distant past with our possible future kin is an ability to survive adverse conditions by splitting into distant but connected bands. And what makes us human is our ability to build homes and communities almost anywhere. We should treasure this skill, because it is the cornerstone of our best survival strategy. We'll strike out into space the way our ancestors once struck out for the world beyond Africa. And eventually we'll evolve into beings suited to our new habitats among the stars.

Things are going to get weird. There may be horrific disasters, and many lives will be lost. But don't worry. As long as we keep exploring, humanity is going to survive.

ACKNOWLEDGMENTS

One of the most amazing parts of writing this book was getting a chance to discuss the future of humanity with so many scientists, engineers, philosophers, historians, technicians, and sundry brainiacs. That over a hundred people would take the time to share their ideas with a stranger, about everything from geological history to space exploration, is testimony to the basic awesomeness of humanity.

Thanks to my fantastic agent, Laurie Fox, for making all of this happen, and to my benevolent editor, Gerald Howard. Thanks to Hannah Wood for tons of editorial help, and to artist Neil Webb (who created the gorgeous cover).

I could not have written this book if my boss, Nick Denton, hadn't given me the time to do it–thanks, Nick! And thanks to the io9 crew for always inspiring me: Charlie Jane Anders, Cyriaque Lamar, Esther Inglis-Arkell, George Dvorsky, Lauren Davis, Meredith Woerner, Robbie Gonzalez, Rob Bricken, and Steph Fox.

Thanks to members of the unnamed writing group: Claire Light, Sacha Arnold, Nicole Gluckstern, Lee Konstantinou, and Naamen Tilahun.

Many friends and strangers read early versions of the manuscript and gave me feedback: Deb Chachra, Tom Levinson, Maggie Koerth-Baker, Ed Yong, Terry Johnson, Dave Goldberg, Matthew Clapham, Peter Eckersley, and Daniel Rokhsar. None of the factual errors in this book are their doing.

Most important, thanks to three wonderful hominins, Charlie, Chris, and Jesse, who put up with all kinds of nonsense while I was writing this book, and whose love makes me the luckiest person in this limb of the galaxy.

NOTES

INTRODUCTION: ARE WE ALL GOING TO DIE?

3 30 percent of bee colonies: See U.S. Department of Agriculture, "Colony Collapse Disorder Progress Report" (2011), in which the Colony Collapse Disorder Steering Committee reports, "Annual surveys clearly show that overall colony losses continue to be as high as 30 percent or more since CCD began to be reported," http://www.ars.usda.gov/is/br/ccd/ccdprogressreport2011.pdf.

3 amphibian crisis: D. B. Wake and V. T. Vredenburg, "Are We in the Midst of the Sixth Mass Extinction? A View from the World of Amphibians," *Proceedings of the National Academy of Sciences* 105 (2008): 11466–73.

3 E. O. Wilson estimates that 27,000 species of all kinds go extinct per year: I should note that Wilson's estimate has been extremely controversial in the conservationist community, with some scientists strongly disagreeing with the way he reached this number. Still, most biologists who disagree with the size of the number do not disagree with the notion that we are seeing a rise in extinctions. The estimate comes from E. O. Wilson, *The Diversity of Life* (Cambridge, MA: Belknap Press, 1992).

3 coined in the 1990s by the paleontologist Richard Leakey: See Richard Leakey, *The Sixth Extinction: Patterns of Life and the Future of Humankind* (New York: Anchor Press, 1996).

4 Elizabeth Kolbert has tirelessly reported on scientific evidence: Elizabeth Kolbert, "The Sixth Extinction?" *The New Yorker* (May 25, 2009): 53.

8 when you meet Earth scientist Mike Benton: Personal interview, November 2010. Previously quoted in my article "How to Survive a Mass Extinction," io9.com (Nov. 29, 2010), http://io9.com/5700371/how-to-survive-a-mass-extinction.

CHAPTER ONE: THE APOCALYPSE THAT BROUGHT US TO LIFE

15 Earth is roughly 4.5 billion years old: For a more detailed account of the origins of life on Earth that I summarize here, see Andrew H. Knoll, *Life on a Young Planet: The First Three Billion Years of Evolution on Earth* (Princeton, NJ: Princeton University Press, 2003), and Jan Zalasiewicz, *The Planet in a Pebble: A Journey into Earth's Deep History* (Oxford: Oxford University Press, 2010).

16 and Roger Summons is one of them: Personal interview, August 22, 2011.

17 asked his student Dawn Sumner: P. F. Hoffman and D. P. Schrag, "The Snowball Earth Hypothesis," *Terra Nova,* vol. 14, no. 3 (2002): 129–55.

18 I visited Kirschvink at the California Institute of Technology: Personal interview, October 11, 2011. You can read Kirschvink's groundbreaking paper on Snowball Earth, "Late Proterozoic Low-Latitude Global Glaciation: The Snowball Earth," in J. W. Schopf and C. Klein, eds., *The Proterozoic Biosphere: A Multidisciplinary Study* (New York: Cambridge University Press, 1992). It's also available online here: http://www.gps.caltech.edu/users/jkirschvink/pdfs/firstsnowball.pdf.

20 Bill McKibben, who argues in his book *Eaarth:* Bill McKibben, *Eaarth: Making Life on a Tough New Planet* (New York: Times Books, 2010).

21 In a remarkable paper published in *Nature:* Anthony Barnosky et al., "Has the Earth's Sixth Mass Extinction Already Arrived?" *Nature* 471 (March 3, 2011): 51–57.

21 The statistician and paleontologist Charles Marshall: Personal interview, October 18, 2011.

CHAPTER TWO: TWO WAYS TO GO EXTINCT

24 Peter M. Sheehan, a geologist with the Milwaukee Public Museum: Peter M. Sheehan et al., "Understanding the Great Ordovician Biodiversification Event (GOBE): Influences of Paleogeography, Paleoclimate, or Paleoecology?" *GSA Today,* v. 19, no. 4/5 (April/May 2009).

27 "We are seeing a mechanism that changed": Young said this through a press release from Ohio State University about his NSF-funded research. See Pam Frost Gorder, "Appalachian Mountains, Carbon Dioxide Caused Long-Ago Global Cooling," *Ohio State University Research News* (October 25, 2006).

28 Adrian Melott, a professor of physics and astronomy at the University of Kansas: Personal interview, September 27, 2011. For more information on the gamma-ray theory and the 63-million-year cycle, you can read A. Melotte et al., "Did a Gamma-Ray Burst Initiate the Late Ordovician Mass Extinction?" *International Journal of Astrobiology* 3 (2004): 55, and Robert A. Rohde and Richard A. Muller, "Cycles in Fossil Diversity," *Nature* 434 (March 10, 2005): 208–10.

30 Donald Canfield conducted a study of the atmosphere: Donald E. Canfield, et al., "Devonian Rise in Atmospheric Oxygen Correlated to the Radiations of Terrestrial Plants and Large Predatory Fish," *Proceedings of the National Academy of Science* 107 (October 19, 2010): 17911–15.

31 Alycia Stigall, has a theory that could explain: Personal interview, September 22, 2011. See also Alycia Stigall, "Invasive Species and Biodiversity Crises: Testing the Link in the Late Devonian," *PLoS One* 5(12) (2010): e15584.

CHAPTER THREE: THE GREAT DYING

34 Paul Renne, the center's head geologist: Personal interview, October 4, 2011. See also Paul Renne et al., "Synchrony and Causal Relations Between Permian-Triassic Boundary Crises and Siberian Flood Volcanism," *Science* 269 (September 8, 1995): 1413–16.

35 Jonathan Payne, a geologist at Stanford: Personal interview, November 7, 2011. See also Payne and his colleagues' paper on this topic: Jonathan L. Payne et al., "Calcium Isotope Constraints on the End-Permian Mass Extinction," *PNAS* 107 (May 11, 2010): 8543–48.

39 The Permian expert Mike Benton: For a terrific account of what happened during the Permian mass extinction, see Michael J. Benton, *When Life Nearly Died: The Greatest Mass Extinction of All Time* (London: Thames & Hudson, 2003).

40 For answers, I visited Peter Roopnarine, a zoologist: Personal interview, November 21, 2011. See also P. D. Roopnarine, "Ecological Modeling of Paleocommunity Food Webs," in G. Dietl and K. Flessa, eds., *Conservation Paleobiology, The Paleontological Society Papers* 15 (2009).

CHAPTER FOUR: WHAT REALLY HAPPENED TO THE DINOSAURS

46 "It is very hard to imagine what happened": Personal interview, February 1, 2012. For more about what might have happened directly after the bolide impact, see Jan Smit et al., "The Aftermath of the Cretaceous-Paleogene Bolide Impact," *Geophysical Research Abstracts* 13 (2011): 12724. And for Smit and his colleagues' original groundbreaking paper about the bolide impact, see Jan Smit et al., "An Extraterrestrial Event at the Cretaceous-Tertiary Boundary," *Nature* 285 (May 22, 1980).

46 Cretaceous-Tertiary (K-T) mass extinction: Though most mass extinctions are usually referred to using the geological periods they ended (such as the Permian mass extinction, or the Ordovician mass extinction), the K-T mass extinction is known by a name that refers to two geological periods, the Cretaceous (which follows the Jurassic) and the Tertiary. These are two names for roughly the same period of geological time, which was ended by a mass extinc-

tion. Welcome to the weirdness of geological nomenclature, which is a confusing mix of older and newer names, some of which refer to overlapping stretches of time. Making things even more complex are the many names for geological periods used in Asia and other regions of the world. Some paleontologists prefer to call this extinction the Cretaceous-Paleogene (K-P) mass extinction, because the Paleogene is the period that comes after the Cretaceous. However, you're more likely to hear the mass extinction called K-T, so I've chosen to call it that here.

49 "iridium anomaly": See the Alvarezes' first paper on this subject here: L. W. Alvarez, W. Alvarez, F. Asaro, and H. V. Michel, "Extraterrestrial Cause for the Cretaceous-Tertiary Extinction," *Science* 208 (1980): 1095–1108.

49 Princeton geologist Gerta Keller began publishing papers: Personal interview, September 23, 2011. See also her papers about the discoveries she made in India: G. Keller et al., "Deccan Volcanism Linked to the Cretaceous-Tertiary Boundary Mass Extinction: New Evidence from ONGC Wells in the Krishna-Godavari Basin," *Journal of the Geological Society of India* 78 (2011): 399–428; and G. Keller et al., "Environmental Effects of Deccan Volcanism Across the Cretaceous–Tertiary Transition in Meghalaya, India," *Earth and Planetary Science Letters* 310 (October 2011): 272–85.

49 UC Berkeley paleontologist Charles Marshall said: Personal interview, October 18, 2011. Marshall made this comment after I asked him what he thought of Keller's theories.

49 Smit told the BBC: This comment comes from an interview Smit did with the BBC program *Horizon,* in the episode "What Really Killed the Dinosaurs?" You can read a transcript here: http://www.bbc.co.uk/sn/tvradio/programmes/horizon/dino_trans.shtml.

50 the dinosaurs died of fungal infections: See Arturo Casadevall, "Fungal Virulence, Vertebrate Endothermy, and Dinosaur Extinction: Is There a Connection?" *Fungal Genetics and Biology* 42 (2005): 98–106.

51 Brown University geologist Jessica Whiteside put it: Personal interview, January 12, 2012. See also: J. H. Whiteside, P. E. Olsen, D. V. Kent, S. J. Fowell, and M. Et-Touhami, "Synchrony Between the CAMP and the Triassic-Jurassic Mass-Extinction Event? Reply to Comment of Marzoli et al," *Palaeogeography, Palaeoclimatology, and Palaeoecology* 262 (2008): 194–98.

51 CAMP: A helpful map of the CAMP eruption can be found here: http://www.auburn.edu/academic/science_math/res_area/geology/camp/Fig1.jpg.

52 Jennifer McElwain, a paleobotanist at University College Dublin: Personal interview, January 16, 2012.

53 dinofuzz: See Ryan C. McKellar et al., "A Diverse Assemblage of Late Cretaceous Dinosaur and Bird Feathers from Canadian Amber," *Science* 16 (2011): 1619–22.

53 they may have been social, like birds: See, for example, D. J. Varricchio, Paul C. Sereno, Zhao Xijin, Tan Lin, Jeffery A. Wilson, and Gabrielle H. Lyon, "Mud-Trapped Herd Captures Evidence of Distinctive Dinosaur Sociality," *Acta Palaeontologica Polonica* 53 (2008): 567–78.

CHAPTER FIVE: IS A MASS EXTINCTION GOING ON RIGHT NOW?

56 University of Oregon archaeologist Dennis Jenkins: Personal interview, February 9, 2012. See also Dennis Jenkins et al., "Clovis Age Western Stemmed Projectile Points and Human Coprolites at the Paisley Caves," *Science* 337 (2012): 223–28.

57 UC Berkeley biologist Anthony Barnosky has been at the forefront: Personal interview, October 18, 2011.

58 March 3, 2011, issue of *Nature:* Anthony Barnosky et al., "Has the Earth's Sixth Mass Extinction Already Arrived?" *Nature* 471 (2011): 51–57.

59 Peter Ward, a geologist at the University of Washington: Personal interview, September 27, 2011. See also Peter Ward, *The Medea Hypothesis: Is Life on Earth Ultimately Self-Destructive?* (Princeton, NJ: Princeton University Press: 2009).

CHAPTER SIX: THE AFRICAN BOTTLENECK

64 "effective population size": For more on this concept, see Matthew B. Hamilton's valuable primer *Population Genetics* (West Sussex: Wiley Blackwell, 2009), plus you can find a good introduction to this idea in R. Kliman, B. Sheehy, and J. Schultz, "Genetic Drift and Effective Population Size," *Nature Education* 1 (2008).

64 human effective population size: N. Takahata, Y. Satta, and J. Klein, "Divergence Time and Population Size in the Lineage Leading to Modern Humans," *Theoretical Population Biology* 48 (October 1995): 198–221.

65 Toba catastrophe: Richard Dawkins, *The Ancestor's Tale, A Pilgrimage to the Dawn of Life* (Boston: Houghton Mifflin Company, 2004).

65 But, as John Hawks, an anthropologist at the University of Wisconsin, Madison, put it to me: Personal interview, November 29, 2011.

65 according to the anthropologist Ian Tattersall of the American Museum of Natural History: Personal interview, December 13, 2011.

65 we were part of a hominin group: Most of the evolutionary history I describe here comes from Richard Klein's indispensable evolutionary biology text *The Human Career: Human Biological and Cultural Origins* (Third Edition) (Chicago: University of Chicago Press, 2009). In this book I'm following today's accepted scientific practice of referring to humans and our ancestors as "hominins." Hominids include the greater group of humans and apes (and their ancestors).

69 But as the Stanford paleoanthropologist Richard Klein told me: Personal interview, November 30, 2011.

71 but the University of Utah geneticist Chad Huff recently argued: Chad Huff et al., "Mobile Elements Reveal Small Population Size in the Ancient Ancestors of Homo Sapiens," *PNAS* 107 (February 2, 2010): 2147–52.

71 As Hawks explained in a paper he published with colleagues in 2000: Hawks et al., "Population Bottlenecks and Pleistocene Human Evolution," *Molecular Biology and Evolution* 17 (2000): 2–22.

71 That's how speciation creates a genetic bottleneck: It's worth noting that the definition of "species" can get as messy as evolution itself. For example, two classes of species may look similar but have very different genetic backgrounds, like bats (from the class Mammalia) and birds (from the class Aves); two species can crossbreed and produce viable offspring, like bonobos and chimps; and two very different-looking animals might actually be the same species at different stages in their development, like tadpoles and frogs. Scientists who study phylogeny, the family trees that define a species, use many rubrics to draw boundaries between species, but there are no absolute rules. But when two groups' genetic makeup, body shape, and behaviors diverge enough, they are generally said to be separate species. There is a great deal of debate over where to draw species lines between ancient human groups, but we can say for certain that today's *H. sapiens* descended from a pretty small genetic pool—and speciation could be part of what made that pool smaller.

72 his book *Prehistoric Art:* Randall White, *Prehistoric Art: The Symbolic Journey of Humankind* (New York: Abrams, 2003).

73 In his book *The Mating Mind:* Geoffrey Miller, *The Mating Mind: How Sexual Choice Shaped the Evolution of Human Nature* (New York: Anchor Books, 2001).

73 *The 10,000 Year Explosion:* Gregory Cochran and Henry Harpending, *The 10,000 Year Explosion: How Civilization Accelerated Human Evolution* (New York: Basic Books, 2010).

73 humans bred themselves to be the ultimate survivors: It seems counterintuitive to say that we're survivors when most of us have been taught that genetic diversity is vital to species health. Does our low genetic diversity mean that we're weaker than other species? In some cases, as we'll see later , it can be a vulnerability. However, the kind of inbreeding that creates low effective population size is not the same thing as inbreeding between close relatives like a brother and sister, which can result in genetic defects. Indeed, as Murdoch University's geneticists Alan Bittles and Michael Black point out in a paper about human "consanguinity," or inbreeding, it was long considered acceptable for second cousins and more distant relatives to marry in Western cultures—and is still common in many parts of the world. Such marriages have few deleterious effects, and as we've seen, these traditions grow out of what is probably a very ancient

human practice in founder populations. Inbreeding between distant relatives seems to be the human norm.

74 DNA extracted from the fossils of Neanderthals and other hominins: Svante Pääbo et al., "A Draft Sequence of the Neanderthal Genome," *Science* 328 (May 7, 2010): 710–22.

CHAPTER SEVEN: MEETING THE NEANDERTHALS

78 Neanderthals used tools and fire: See Brian Fagan, *Cro-Magnon: How the Ice Age Gave Birth to the First Modern Humans* (New York: Bloomsbury Press, 2011).

78 Neanderthals were mostly meat-eaters: Fagan, *Cro-Magnon.*

78 Richard Klein: Personal interview, November 30, 2011.

79 possibly with red hair: The possibility of some Neanderthals having pale skin and red hair is raised in Carles Lalueza-Fox et al., "A Melanocortin 1 Receptor Allele Suggests Varying Pigmentation Among Neanderthals," *Science* 318 (2007): 1453–55.

79 Many Neanderthal skeletons are distorted by broken bones that healed: Joe Alper, "Rethinking Neanderthals," *Smithsonian Magazine* (June 2003).

79 a generous estimate: John Hawks, personal correspondence, October 9, 2012.

79 many scientists believe: V. Fabre, S. Condemi, and A. Degioanni, "Genetic Evidence of Geographical Groups Among Neanderthals," *PLoS ONE* 4 (2009): e5151.

80 A 60,000-year-old Neanderthal grave: Jennifer Viegas, "Did Neanderthals Believe in an Afterlife?" *Discovery News* (April 20, 2011).

80 Neanderthals talked or even sang: See Steven Mithen, *The Singing Neanderthals* (Cambridge: Harvard University Press, 2007).

82 two dominant theories: John Relethford, "Genetics of Modern Human Origins and Diversity," *Annual Review of Anthropology* 27, (1998): 1–23.

82 Rebecca Cann and her colleagues found a way to support: Rebecca L. Cann et al., "Microbial DNA and Human Evolution," *Nature* 325 (1987): 31–36.

83 Popularized by Wolpoff and his colleague John Hawks: See Milford Wolpoff and John Hawks, "Modern Human Origins," *Science* 241 (August 12, 1988): 772–74.

84 "out of Africa" migration is based on an artificial political boundary: According to Clive Finlayson, who writes, "The first proto-humans would have gradually expanded into favorable habitats wherever these were." Clive Finlayson, *The Humans Who Went Extinct* (Oxford: Oxford University Press, 2009).

84 Hawks has presented compelling genetic evidence: John Hawks, Gregory Cochran, Henry C. Harpending, and Bruce T. Lahn, "A Genetic Legacy from Archaic Homo," *Trends in Genetics* 24 (January 2008): 19–23.

84 the truth lies somewhere in between African replacement and multiregionalism: The middle-of-the-road take on all this is often dubbed the "assimilation theory." Anthropologist Vinayak Eswaran (see Vinayak Eswaran et al., "Genomics Refutes an Exclusively African Origin of Humans," *Journal of Human Evolution* 49 [July 2005]: 1–18) and his colleagues take this perspective in a paper in which they argue that genetic evidence suggests that there were two distinct waves of immigration out of Africa—the archaic human one and the *H. sapiens* one. But as *H. sapiens* moved out into the world, they assimilated the local Neanderthal peoples, along with their other cousin *H. erectus* in Asia.

So basically, in the assimilation-theory model, *H. sapiens* didn't destroy their kindred, nor were they deeply interrelated with them as in the multiregional theory. They met them as strangers, but forged alliances and formed families with them. Gradually, though, *H. sapiens* became the dominant culture.

87 Simon Armitage published a paper suggesting that *H. sapiens* emerged from Africa: Simon Armitage et al., "The Southern Route 'Out of Africa': Evidence for an Early Expansion of Modern Humans into Arabia," *Science* 331 (2011): 453–56.

87 Svante Pääbo, who led the Neanderthal DNA sequencing project: Svante Pääbo et al., "A High-Coverage Genome Sequence from an Archaic Denisovan Individual," *Science* 338 (October 2012): 222–26.

CHAPTER EIGHT: GREAT PLAGUES

89 wiped out over 60 percent of the population of the British Isles: Ole Jørgen Benedictow, *The Black Death, 1346–1353: The Complete History* (Woodbridge, U.K.: The Boydell Press, 2004).

89 The son of a wealthy wine merchant, Chaucer grew up: All the biographical details here come from Larry Benson, Robert Pratt, and F. N. Robinson, eds., *The Riverside Chaucer* (New York: Houghton Mifflin, 1986).

90 our own growing societies: We see the first stirrings of modern global culture during the late Middle Ages in Europe, when a growing middle class began laying the foundations for capitalism and global trade communities. As the sociologist Anthony Giddens would have it, this was the moment when the premodern era gave way to the modern. In a sense, today's world is part of a narrative arc that began during Chaucer's time. Before that era, urban or global cultures still tended to be exceptions.

Probably the closest historical analogy to the global culture that began stirring to life during Chaucer's time would have been that of the Silk Road trade route that crossed from China into Europe for over a millennium. Still, the Silk Road culture was also a localized phenomenon, accessible only to nearby regions. Within a few hundred years of Chaucer's time, the oceans became in essence an enormous Silk Road, uniting every continent in the world.

92 Jo Hays, a historian at Loyola University whose work focuses on pandemics: Personal interview, October 2011.

93 Likewise, the common people began questioning government authorities: Robert S. Gottfried, *The Black Death: Natural and Human Disaster in Medieval Europe* (New York: The Free Press, 1983), especially chapter 5.

93 After the Black Death, there was a rise: Frances and Joseph Gies, *Women in the Middle Ages* (New York: Harper Perennial, 1991).

94 SUNY Albany anthropologist Sharon DeWitte: Personal interview, November 30, 2011.

94 sequenced bacterial DNA: Kirsten I. Bos et al., "A Draft Genome of *Yersinia pestis* from Victims of the Black Death," *Nature* 478 (October 27, 2011): 506–10.

96 New York University's literary historian Ernest Gilman: Personal interview, February 15, 2012. See also Ernest Gilman, *Plague Writing in Early Modern England* (Chicago: University of Chicago Press, 2009).

98 In the late 1990s, Jared Diamond argued: Jared Diamond, *Guns, Germs and Steel: The Fates of Human Societies* (New York: W. W. Norton & Co., 1999, originally published in hardback in 1997). Diamond's argument about the Americas is contained in a couple of chapters where he discusses the conquest of the massive Incan army by Pizarro's small band of hooligans. Though very much aware of how plagues also played into this scenario, Diamond focuses in these chapters most heavily on how the Inca lacked steel and writing.

98 As Charles Mann explains in his book *1491:* See Charles Mann, *1491: The Revelations of the Americas Before Columbus* (New York: Vintage, 2005), especially part one, where Mann discusses how historians have arrived at the 90 percent number I mention a few paragraphs later.

98 only today finally being deciphered: One of the main places this decipherment is being done is at the Khipu Database Project at Harvard University. Scholars involved have identified several ways that the knots convey meaning, including size, orientation, color, and shape. They've already figured out the numbering system and are now moving on to the written language. You can learn more about the project here: http://khipukamayuq.fas.harvard.edu/.

99 Arizona State University forensic archaeologist Jane Buikstra: Personal interview, February 14, 2012.

99 Paul Kelton, of the University of Kansas: Personal interview, February 2012. See also Paul Kelton, *Epidemics and Enslavement: Biological Catastrophe in the Native Southeast 1492–1715* (Lincoln and London: University of Nebraska Press, 2007).

100 David S. Jones, a Harvard historian and medical doctor, sums up the

issues: David S. Jones, "Virgin Soils Revisited," *William and Mary Quarterly* 60 (October 2003): 703–42.

101 Susan Kent explains in her recent book about the 1918–19 flu epidemic: Susan Kent, *The Influenza Pandemic of 1918–1919* (Boston and New York: Bedford/St. Martins, 2013).

102 native cultures and peoples have survived throughout the Americas: Gerald Vizenor, *Native Liberty: Natural Reason and Cultural Survivance* (Lincoln and London: University of Nebraska Press, 2009).

CHAPTER NINE: THE HUNGRY GENERATIONS

103 Famines have been recorded in historical documents: Cormac Ó Gráda, *Famine: A Short History* (Princeton and Oxford: Princeton University Press, 2009).

103 Cormac Ó Gráda has spent most of his career: Personal interview, December 8, 2011.

104 This famine had its roots in politics: John O'Rourke, *The History of the Great Irish Famine of 1847* (Dublin: James Duffy and Co., Ltd., 1902). This is a fascinating account by a man who gathered together many first-person accounts of the famine, mostly from interviews he did with survivors in the late nineteenth century. He credits "the public press" as being one of the first groups to alert the world to the famine, and place it in a political context.

104 Amartya Sen first advanced this theory in the 1980s: Amartya Sen, *Poverty and Famines: An Essay on Entitlement and Deprivation* (Oxford: Oxford University Press, 1983).

105 Evan Fraser, a geographer at the University of Guelph: Personal interview, February 16, 2012. See also Evan Fraser, "Social Vulnerability and Ecological Fragility: Building Bridges Between Social and Natural Sciences Using the Irish Potato Famine as a Case Study," *Conservation Ecology* 7 (2003): 9.

106 According to the weekly U.S. Drought Monitor: I've pulled these numbers from publicly available statistics on the summer 2012 Midwest drought, available from the U.S. National Climatic Data Center in its August 2012 State of the Climate drought report. This document is available online here: http://www.ncdc.noaa.gov/sotc/drought/#national-overview.

107 Newcastle University historian and demographer Violetta Hionidou: Personal interview, February 15, 2012. See also Violetta Hionidou, *Famine and Death in Occupied Greece, 1941–1944* (Cambridge, U.K.: Cambridge University Press, 2006).

109 University of Hong Kong history professor Frank Dikötter: Frank Dikötter, *Mao's Great Famine: The History of China's Most Devastating Catastrophe, 1958–1962* (New York: Walker and Company, 2011).

111 Intergovernmental Panel on Climate Change's recent models: You can see these predictions in S. Solomon, D. Qin, M. Manning, Z. Chen, M. Marquis, K. B. Averyt, M. Tignor, and H. L. Miller, eds.,

Contribution of Working Group I to the Fourth Assessment Report of the Intergovernmental Panel on Climate Change, 2007 (Cambridge: Cambridge University Press, 2007). It is online at http://www.ipcc.ch/publications_and_data/ar4/wg1/en/spmsspm-projections-of.html.

112 "You get a famine if the price of food": Personal correspondence, February 14, 2012.

CHAPTER TEN: SCATTER: FOOTPRINTS OF THE DIASPORA

119 Stories about how cool it is to rip: Yes, I'm joking a little bit here—I don't think anybody actually talks about ripping faces off in the Old Testament, but there is a lot of chopping off of various body parts and driving stakes through people's heads and similarly graphic violence against enemies. In Assyrian cuneiform tablets, which were often erected as ceremonial stelae or monoliths in celebration of various kings, it was standard practice to praise the current leader by recounting all his battle victories. Indeed, we get some of the first historical accounts of the Jewish people in one of these stelae, in the Louvre's collection. On it, King Sargon talks about how great it was to capture and kill thousands of Jews from the northern kingdom of Israel, called Samaria. It's important to remember that this kind of writing was part of the style of national monuments of the era, and probably didn't reflect the common people's sentiments or even the sentiments of the people writing. They were patriotic documents, intended as propaganda. But it was against the backdrop of this kind of propaganda that Exodus was written and compiled, which makes many aspects of the story quite remarkable.

119 In modern parlance, the term "diaspora": William Safran, "Diasporas in Modern Societies: Myths of Homeland and Return," *Diaspora: A Journal of Transnational Studies* 1 (1991). See also Robin Cohen, *Global Diasporas: An Introduction—Second Edition* (New York: Routledge, 2008).

120 UC Berkeley archaeologist Carol Redmount: "Bitter Lives: Israel In and Out of Egypt," from *The Oxford History of the Biblical World* (Oxford: University of Oxford Press, 1998).

120 But then in the eighth century: Israel Finkelstein and Neil Asher Silberman, *The Bible Unearthed: Archaeology's New Vision of Ancient Israel and the Origin of Its Sacred Texts* (New York: The Free Press, 2001).

121 adopting the local language, Aramaic: "Into Exile: From the Assyrian Conquest of Israel to the Fall of Babylon," Mordechai Cogan, from *The Oxford History of the Biblical World* (Oxford: University of Oxford Press, 1998).

121 Yehudim, or Jews: Ibid.

122 As geneticist David Goldstein notes in his book: David B. Goldstein, *Jacob's Legacy: A Genetic View of Jewish History* (New Haven and London: Yale University Press, 2008).

122 we know from contemporary sources: Leonard Victor Rutgers,

"Roman Policy Towards the Jews: Expulsions from the City of Rome During the First Century C.E.," *Classical Antiquity*, vol. 13, no. 1 (April 1994): 56–74.

123 Ostrer wanted to know: Personal interview, April 6, 2012.

123 researchers at the Jewish HapMap project scoured their data: See Gil Atzmon, Li Hao, Itsik Pe'er, Christopher Velez, Alexander Pearlman et al., "Abraham's Children in the Genome Era: Major Jewish Diaspora Populations Comprise Distinct Genetic Clusters with Shared Middle Eastern Ancestry," *The American Journal of Human Genetics* 86 (June 11, 2010): 850–59. You can also read Harry Ostrer's popular account of their work in *Legacy: A Genetic History of the Jewish People* (Oxford: Oxford University Press, 2012).

124 hints about where people's ancestors settled in the diaspora: Scientists can even narrow down the time period when different groups likely split up and headed in different directions. Earlier in this book, we talked about how evolutionary biologists tracking the origins of *Homo sapiens* can trace the divergence of two species by looking at DNA shared between them and assuming a fixed rate of mutation, or change over time. The divergence of two or more haplotypes can be traced the same way. Syrian Jews and Eastern European Jews, for example, share many long strands of DNA. But in the centuries since those two groups split apart, those strands have accumulated a lot of random mutations. Assuming a fixed rate of mutation over time, scientists like Ostrer can estimate that the two groups likely split up roughly 2,500 years ago. And, by looking at the geographical distribution of haplotypes over Europe, some scientists have even started to track migration paths. See W. Y. Yang, J. Novembre, E. Eskin, and E. Halperin, "A Model-Based Approach for Analysis of Spatial Structure in Genetic Data," *Nature Genetics* 44 (2012): 725–31.

124 once again sent Jews running: Henry Kamen, *The Spanish Inquisition: A Historical Revision* (New Haven: Yale University Press, 1999).

124 A group of Portuguese anthropologists: Inês Nogueiro, Licínio Manco, Verónica Gomes, António Amorim, and Leonor Gusmão, "Phylogeographic Analysis of Paternal Lineages in NE Portuguese Jewish Communities," *American Journal of Physical Anthropology* 141 (March 2010): 373–81.

125 *The Black Atlantic: Modernity and Double-Consciousness:* Paul Gilroy, *The Black Atlantic* (Reissued Edition) (Boston: Harvard University Press, 1993).

CHAPTER ELEVEN: ADAPT: MEET THE TOUGHEST MICROBES IN THE WORLD

127 Its subsequent 3.5-billion-year career: T. N. Taylor and E. L. Taylor, *The Biology and Evolution of Fossil Plants* (Upper Saddle River, NJ: Prentice Hall, 1993).

127 cyano evolved inside other cells: This account of plant-cell evolution

is called endosymbiotic theory, and originated over a century ago. Today it's fairly widely accepted and is backed by genetic evidence. See, for example, Geoffrey I. McFadden and Giel G. van Dooren, "Evolution: Red Algal Genome Affirms a Common Origin of All Plastids," *Current Biology* 14 (July 13, 2004): R514–16.

128 Brett Neilan, a biologist at the University of New South Wales: Personal interview, January 15, 2012.

128 circadian rhythms of light and dark: Hideo Iwasaki and Takao Kondo, "The Current State and Problems of Circadian Clock Studies in Cyanobacteria," *Plant Cell Physiology* 41 (2000): 1013–20.

129 ubiquitous and sustainable: At least until the Sun incinerates the Earth in about a billion years.

129 One of these scientists is physicist-turned-biologist Himadri Pakrasi: Personal interviews, January 6 and March 8, 2012.

131 "You know why most plants are green?": Personal interview, March 9, 2012.

132 Environmental engineer Richard Axelbaum: Personal interview, March 8, 2012.

132 As a result, the only by-products: S. A. Skeen, B. M. Kumfer, and R. L. Axelbaum, "Nitric Oxide Emissions During Coal and Coal/Biomass Combustion Under Air-fired and Oxy-fuel Conditions," *Energy & Fuels* 24 (2010): 4144–52.

133 his team made an incredible breakthrough: Anindita Bandyopadhyay, Jana Stöckel, Hongtao Min, Louis A. Sherman, and Himadri B. Pakrasi, "High Rates of Photobiological H2 Production by a Cyanobacterium Under Aerobic Conditions," *Nature Communications* 1 (December 14, 2010).

133 Steven Chu has talked about replacing the oil economy: "The Alternative Choice: Steven Chu Wants to Save the World by Transforming Its Largest Industry: Energy," *The Economist* (July 2, 2009).

CHAPTER TWELVE: REMEMBER: SWIM SOUTH

135 Charles Melville Scammon wrote about his experiences hunting grays: Charles Melville Scammon, *The Marine Mammals of the North-Western Coast of North America* (San Francisco: JH Carmany, and New York: Putnam, 1874). Full text available via the Internet Archive at http://archive.org/details/marinemammalsofn00scam.

136 Their brains "sleep": Most scientists believe that gray whales, like other cetaceans, experience "unihemispheric slow wave sleep," where only one hemisphere of the brain "sleeps" at a time. Whales also experience little to no REM sleep. See, for example, Oleg I. Lyamin, Paul R. Manger, Sam H. Ridgway, Lev M. Mukhametov, and Jerome M. Siegel, "Cetacean Sleep: An Unusual Form of Mammalian Sleep," *Neuroscience & Biobehavioral Reviews* 32 (October 2008): 1451–84.

137 2.5 million years ago: N. D. Pyenson and D. R. Lindberg, "What Happened to Gray Whales during the Pleistocene? The Ecological

Impact of Sea-Level Change on Benthic Feeding Areas in the North Pacific Ocean," *PLoS ONE* 6 (2011): e21295.

138 Harvey, now a professor at Moss Landing Marine Laboratories: Personal interview, February 2, 2012. All subsequent quotes from Harvey are from this interview.

140 grays are migrating later in the year: John Upton, "Scientists Look Far to the North to Explain Young Whale in San Francisco Bay," *New York Times* (March 17, 2012).

140 got caught in the frozen Artic in 1988: This story is recounted by the journalist Tom Rose, in his book *Big Miracle* (Reprint) (New York: St. Martins, 2011). Originally published in 1989, under the title *Freeing the Whales: How the Media Created the World's Greatest Non-Event.*

141 A large group of grays lived in the Atlantic for thousands of years: Scott Noakes, "Georgia's Pleistocene Atlantic Gray Whales," *Gray's Reef National Marine Sanctuary,* http://graysreef.noaa.gov/science/research/gray_whale/welcome.html.

141 20,000–30,000 individuals: This number is a point of some debate, as populations change every year, and the only way we can count them is by observing how many different individuals pass by observation stations along the Pacific Coast. The number that scientists seem to agree on most is 22,000 individuals, based on average numbers from data collected over the past two decades. The population is growing by over 2 percent every year, and in 2012 over 200 babies were born. I'm basing my numbers on two sources: S. Elizabeth Alter, Eric Rynes, and Stephen Palumbi, "DNA Evidence for Historic Population Size and Past Ecosystem Impacts of Gray Whales," *PNAS* 104 (September 18, 2007): 15162–67, and the annual population estimate reports issued by NOAA, which collects data from several observation stations along the coast from Alaska to Mexico (http://www.nmfs.noaa.gov/pr/sars/).

141 Korean-Okhotsk grays: O. Yu. Tyurneva, Yu. M. Yakovlev, V. V. Vertyankin, and N. I. Selin, "The Peculiarities of Foraging Migrations of the Korean-Okhotsk Gray Whale (*Eschrichtius robustus*) Population in Russian Waters of the Far Eastern Seas," *Russian Journal of Marine Biology* 36 (March 2010): 117–24.

142 in 1949 the newly formed International Whaling Commission: The IWC was established in 1946, and included several member states such as the United States, Japan, and the then Soviet Union. You can see a record of every annual meeting online (http://www.iwcoffice.org/meetings/historical.htm). What's interesting about this and other early conservation groups is that it combined the interests of environmentalism with commercial interests. It's also worth noting that some Inuit groups are still permitted to hunt a small number of grays every year.

142 Others argue, based on genetic data: Alter et al., "DNA Evidence for Historic Population Size."

142 Several studies suggest that noise pollution: Many of these studies

are collected in Scott D. Kraus and Rosalind M. Rolland, eds., *The Urban Whale: North Atlantic Whales at the Crossroads* (Boston: Harvard University Press, 2007).

CHAPTER THIRTEEN: PRAGMATIC OPTIMISM, OR STORIES OF SURVIVAL

146 "To try to foretell the future": Octavia Butler, "A Few Rules for Predicting the Future," *Essence* (May 2000).

146 Butler recalled visitors making casually racist remarks: Octavia Butler, "Octavia Butler's Aha! Moment," *O, The Oprah Magazine* (May 2002).

146 *Devil Girl from Mars:* Octavia Butler, "*Devil Girl from Mars*: Why I Write Science Fiction," *MIT Communications Forum*, http://web.mit.edu/comm-forum/papers/butler.html.

148 In her trilogy of novels called *Lilith's Brood:* Octavia Butler, *Lilith's Brood* (New York: Grand Central Publishing, 2000). Original trilogy of novels published in 1987, '88, and '89.

151 *Parable of the Sower* and its sequel, *Parable of the Talents:* Octavia Butler, *Parable of the Sower* (New York: Grand Central Publishing, 2000). Originally published in 1993. Octavia Butler, *Parable of the Talents* (New York: Grand Central Publishing, 2000). Originally published in 1998.

152 "I used to despise religion": "Octavia Butler: Persistence," *Locus* (June 2000).

153 "There's no single answer": Octavia Butler, "A Few Rules for Predicting the Future."

CHAPTER FOURTEEN: THE MUTATING METROPOLIS

158 In the past decade, the number of people on Earth living in cities: The highlights of "World Urbanization Prospects" (last revised April 2012), a U.N. Report on population demographics in cities, gives us this statistic, adding that "urban dwellers will likely account for 86 percent of the population in the more developed regions and for 64 percent of that in the less developed regions." You can view the U.N.'s World Urbanization Prospects report, and the data that informs it, on the U.N. website at http://esa.un.org/unpd/wup/index.htm.

158 *The World Without Us:* Alan Weisman, *The World Without Us* (New York: Thomas Dunne Books, 2007).

158 It would require us to regulate the bodies of billions of women: I am talking here about what it would take to lower the population over the next half century. Many studies have shown that birth rates plummet dramatically in countries where women receive the same educational and economic opportunities as men. It is my fervent hope that over the long term, women's equality with men around the world will lead to a population size that is better adapted to the

Earth's environment. Until that happens, however, we must accept that our population is growing and prepare for it.

158 Jane Jacobs, in her groundbreaking 1961 book: Jane Jacobs, *The Death and Life of Great American Cities* (New York: Random House, 1961).

158 Some call it an emergent property: See, for example, Steven Johnson, *Emergence: The Connected Lives of Ants, Brains, Cities, and Software* (New York: Scribner, 2002).

159 And the fantasy author Fritz Leiber dubbed it "megapolisomancy": This is from Fritz Leiber's incredible 1977 urban fantasy novella about San Francisco, *Our Lady of Darkness.*

159 "battle suits for surviving the future": Matt Jones, "The City Is a Battle Suit for Surviving the Future," io9.com (September 20, 2009), http://io9.com/5362912/the-city-is-a-battlesuit-for-surviving-the-future.

159 Anthropologist Monica L. Smith, who researches the development of cities: Monica L. Smith, ed., "Introduction," *The Social Construction of Ancient Cities* (Washington, D.C.: Smithsonian Books, 2003).

159 Spiro Kostof suggested the same thing: Spiro Kostof, *The City Shaped: Urban Patterns and Meanings Through History* (Boston and London: Little, Brown and Company, 1991).

160 Cities were born in two very different regions of the world: A helpful primer on ancient Peruvian cities can be found in Kimberly Munro, "Ancient Peru: The First Cities," *Popular Archaeology* 3 (March 18, 2011). And my information about ancient Mesopotamian cities and their relationship to agriculture comes from Charles Gates, *Ancient Cities: The Archaeology of Urban Life in the Ancient Near East and Egypt, Greece, and Rome* (Second Edition) (London and New York: Routledge, 2011).

162 Ian Hodder, who has led excavations at Çatalhöyük: Ian Hodder and Craig Cessford, "Daily Practice and Social Memory at Çatalhöyük," *American Antiquity* 69 (January 2004).

162 Anthropologist Elizabeth Stone has been excavating ancient cities: Personal interview, April 25, 2012.

163 The differences between ancient and medieval cities are just as stark: For much more nuanced representations of historical urban development, see Kostof, *The City Shaped*; Josef W. Konvitz, *The Urban Millennium: The City-Building Process from the Early Middle Ages to the Present* (Carbondale and Edwardsville: Southern Illinois University Press, 1985); and Richard T. LeGates and Frederic Stout, eds., *The City Reader* (New York and London: Routledge, 1996).

164 As Harvard economist Edward Glaeser puts it in his book: Edward Glaeser, *The Triumph of the City: How Our Greatest Invention Makes Us Richer, Smarter, Greener, Healthier, and Happier* (New York: The Penguin Press, 2011).

165 Urban geographer Richard Walker believes the San Francisco Bay Area provides: Personal interview, June 16, 2012. For Walker's insights into San Francisco as an environmental city, see also his

book *The Country in the City: The Greening of the San Francisco Bay* (Seattle and London: University of Washington Press, 2007).

CHAPTER FIFTEEN: DISASTER SCIENCE

169 experimental reconnaissance robots: See, for example, "Robots Converge on Disaster City," *Disaster Preparedness and Response: TEEX* (March 22, 2010), http://www.teex.org/teex.cfm?pageid=USARresc&area=USAR&storyid=984&templateid=23.

169 At Oregon State's tsunami lab: Officially called the O. H. Hinsdale Wave Research Laboratory, the facility makes its data publicly available so that other researchers can build simulations based on what other scientists have learned about wave behaviors. A good example of how this data-sharing works can be found in this paper: T. E. Baldock, D. Cox, T. Maddux, J. Killian, and L. Fayler, "Kinematics of Breaking Tsunami Wavefronts: A Data Set from Large Scale Laboratory Experiments," *Coastal Engineering* 56 (May–June 2009).

171 I met the UC Berkeley civil engineer Shakhzod Takhirov inside a three-story warehouse: Personal interview, February 16, 2012.

175 Richard Iverson has created hundreds of landslides to learn more: Personal interview, June 26, 2012. You can also see an incredible collection of videos from Iverson's experiments on the USGS Debris Flow Flume site: http://pubs.usgs.gov/of/2007/1315/.

177 George Thomas, a former structural engineer: All comments are from his public presentation "Smarter Cities," at Washington University in St. Louis (March 7, 2012).

178 Japan was unprepared for the calamity: Emily Rauhala, "How Japan Became a Leader in Disaster Preparation," *Time* (March 11, 2011).

CHAPTER SIXTEEN: USING MATH TO STOP A PANDEMIC

182 David Blythe manages health surveillance for the Maryland public-health department: Personal interview, January 26, 2012.

183 "In this classic urban-plague scenario": N. C. Stenseth, B. B. Atshabar, M. Begon, S. R. Belmain, E. Bertherat et al., "Plague: Past, Present, and Future," *PLoS Medicine* 5 (2008): e3.

185 In its report, the WHO speculated: "WHO Issues Consensus Document on the Epidemiology of SARS" (October 17, 2003), http://www.who.int/csr/sars/archive/epiconsensus/en/.

185 Tini Garske is a mathematician and researcher with the Imperial College London's: T. Garske, H. Yu, Z. Peng, M. Ye, H. Zhou et al., "Travel Patterns in China," *PLoS ONE* 6 (2011): e16364.

186 "SARS quarantine in Toronto was both inefficient and ineffective": Richard Schabas, "Severe Acute Respiratory Syndrome: Did Quarantine Help?" *Canadian Journal of Infectious Diseases and Medical Microbiology* 15 (July–August 2004): 204.

186 Brian Coburn, and his colleagues claim that school closures: Brian

J. Coburn, Bradley G. Wagner, and Sally Blower, "Modeling Influenza Epidemics and Pandemics: Insights into the Future of Swine Flu (H1N1)," *BMC Medicine* 7 (2009): 30.

187 Laura Matrajt, a mathematician at the University of Washington in Seattle: L. Matrajt and I. M. Longini, Jr., "Optimizing Vaccine Allocation at Different Points in Time during an Epidemic," *PLoS ONE* 5 (2010): e13767.

188 Dr. Tadataka Yamada of the Bill & Melinda Gates Foundation: Tadataka Yamada, "Poverty, Wealth, and Access to Pandemic Influenza Vaccines," *New England Journal of Medicine* 361 (2009):1129–31.

189 Robert Moss is an immunization researcher: R. Moss, J. M. McCaw, and J. McVernon, "Diagnosis and Antiviral Intervention Strategies for Mitigating an Influenza Epidemic," *PLoS One* 6 (February 4, 2011): e14505.

190 Joseph Wu says his models show that countries should always "hedge": J. T. Wu, A. Ho, E.S.K. Ma, C. K. Lee, D.K.W. Chu et al., "Estimating Infection Attack Rates and Severity in Real Time during an Influenza Pandemic: Analysis of Serial Cross-Sectional Serologic Surveillance Data," *PLoS Medicine* 8 (2011): e1001103.

CHAPTER SEVENTEEN: CITIES THAT HIDE

192 frying the ozone layer off: I'm getting my account of the effects of a gamma-ray burst from astronomer Phil Plait's excellent chapter on the subject in *Death from the Skies! These Are the Ways the World Will End . . .* (New York: Viking, 2008).

193 NORAD (North American Aerospace Defense Command) complex beneath Colorado's Cheyenne Mountain: These days, NORAD has been relocated to another facility and the underground city is run by a skeleton crew. Nuclear weapons technology has advanced enough that the city would likely not survive a direct attack.

194 raids from neighboring groups and from Muslims during the Crusades: See a historical account of the region in Spiro Kostof, *Caves of God: The Monastic Environment of Byzantine Cappadocia* (Cambridge, MA: MIT Press, 1972).

197 a layer of concrete can provide more safety still: For an interesting discussion of the history of designs for radiation shielding, including contemporary thinking on the topic, see J. Kenneth Shultis and Richard E. Faw, "Radiation Shielding Technology," *Health Physics* 88 (June 2005): 587–612. They point out that concrete is such a good material for most kinds of radiation shielding that it is one of the most widely studied substances for this purpose.

197 John Zacharias, a city-planning professor at Montréal's Concordia University: Personal interview, June 5, 2012.

197 "stimulating, varied environments": Raymond Sterling and John Carmody, *Underground Space Design* (New York: Wiley, 1993).

198 Agust Gudmundsson, a geology professor at Royal Holloway, University of London: Personal interview, June 26, 2012.

199 Dmitris Kaliampakos and Andreas Bernardos, two engineers who specialize in underground development: D. Kaliampakos and A. Bernardos, "Underground Space Development: Setting Modern Strategies," *WIT Transactions on the Built Environment* 102 (2008).

199 RÉSO-like structures such as the one: You can see the full plans for this underground city, to be called Amfora, on Zwarts & Jansma Architects' website: http://www.zwarts.jansma.nl/page/1597/en.

200 "Smoke—especially black, sooty smoke": Alan Robock, "New Models Confirm Nuclear Winter," *Bulletin of the Atomic Scientists* 68 (September 1989): 66–74. Not much has changed since Robock published this article in terms of our understanding of how nuclear winter would work. Many climate scientists agree that a massive explosion would plunge the planet into unseasonable winter for at least a year. Interestingly, it's likely a megavolcano would cause the same problems we face in a nuclear disaster—minus the radiation danger. See Alan Robock, "New START, Eyjafjallajökull, and Nuclear Winter," *Eos* 91 (2010).

CHAPTER EIGHTEEN: EVERY SURFACE A FARM

204 Raquel Pinderhughes, an urban planning professor at San Francisco State: Raquel Pinderhughes et al. offer a short account of the history of urban farming in Cuba in an essay called "Urban Agriculture in Havana, Cuba," in *Down to Earth* (New Delhi: Centre for Science and the Environment, 2001).

204 "small-plot intensive farming" (SPIN): You can learn more about SPIN, and view a number of articles and videos about the system, on its website here: http://spinfarming.com/whatsSpin/.

205 skyscraper farms: See Dickson Despommier, *The Vertical Farm: Feeding the World in the 21st Century* (New York: Thomas Dunne Books, 2010).

205 A popular way to transform cities in Germany is by building green roofs: Not likely to be used as a food supply: Khandaker M. Shariful Islam, "Rooftop Gardening as a Strategy of Urban Agriculture for Food Security: The Case of Dhaka City, Bangladesh," *Acta Horticulturae* 643 (2004).

Might help with reduction in cost of cooling buildings: S. Gaffin et al., "Energy Balance Modeling Applied to a Comparison of White and Green Roof Cooling Efficiency and Cool Surfaces and Shade Trees to Reduce Energy Use and Improve Air Quality in Urban Areas," *Greening Rooftops for Sustainable Communities Proceedings* (Washington, D.C.: Green Roofs for Healthy Cities, 2005).

Storm-water runoff: Doug Hutchinson, Peter Abrams, Ryan Retzlaff, and Tom Liptan, "Stormwater Monitoring Two Ecoroofs in Portland, Oregon, USA," proceedings for Greening Rooftops for Sustainable Communities Conference (Chicago: Green Roofs for Healthy Cities, 2003).

205 without burning coal: One of the big issues with variable power sources like solar and wind is storage. For an excellent and thorough treatment of how we could transition our energy infrastructure over to systems that rely partly on variable power, see Maggie Koerth-Baker's excellent book *Before the Lights Go Out: Conquering the Energy Crisis Before It Conquers Us* (Hoboken, NJ: Wiley and Sons, 2012).

206 MIT's environmental-policy professor Judith Layzer: Personal interview, August 22, 2011.

206 vulnerable to the vicissitudes of climate: One of the issues here is obviously the current trend toward a general warming of the global climate. Cities located in fertile areas could, in just a century, find themselves trapped in arid, drought-racked landscapes. This is something that land planners are already worried about, and are preparing for today. How do we build cities and farms that are ready for radical climate change? The environmental journalist Mark Hertsgaard deals with this extensively in *Hot: Living Through the Next Fifty Years on Earth* (New York: Houghton Mifflin Harcourt, 2011).

206 Amy McNally, a geography researcher with the group: Personal interview, February 27, 2012.

207 But it's possible that our political priorities: How would we transform the political and social landscape to prioritize environmental concerns? This is an enormous question that is outside the scope of this book, but luckily many other thinkers have tackled it, including Judith Layzer in *The Environmental Case: Translating Values into Policy* (Third Edition) (Thousand Oaks, CA: CQ Press College, 2011).

208 New York architect David Benjamin: Personal interview, April 5, 2012. For more about his AutoCAD-like software for biological design, you can look at the project website: http://www.autodesk research.com/projects/biocompevolution.

209 The students described BacillaFilla: For more about this substance, you can see the BacillaFilla project page here: http://2010.igem.org/ Team:Newcastle.

209 synthetic-biology designer Rachel Armstrong's: Personal interview, May 5, 2012.

210 who hope to use experimental proto-cells: Armstrong describes some of her work on the Venice reef in her e-book *Living Architecture: How Synthetic Biology Can Remake Our Cities and Reshape Our Lives* (TED Books, 2012).

CHAPTER NINETEEN: TERRAFORMING EARTH

216 Bill McKibben and Mark Hertsgaard: See, for example, McKibben's books *The End of Nature* (New York: Random House, 1989) and *Eaarth: Making a Life on a Tough New Planet* (New York: Times

Books, 2010); and Hertsgaard's *Hot: Living Through the Next 50 Years on Earth* (New York: Houghton Mifflin Harcourt, 2011).

216 Maggie Koerth-Baker points out: Maggie Koerth-Baker, *Before the Lights Go Out: Conquering the Energy Crisis Before It Conquers Us* (New York: Wiley, 2012).

216 Government representatives who attend the annual U.N. Climate Change Conferences: Currently, our most urgent task as a planet is to form international agreements that limit carbon emissions. The only certainty when it comes to climate change is that if we limit fossil-fuel use we will slow down the warming process that threatens to cause food shortages and the sixth mass extinction. We already have the technological ability to reduce carbon emissions. How we will do it politically and socially is outside the scope of this book, though it's likely that the same international bodies regulating emissions will ultimately regulate any geoengineering projects we undertake.

217 "hack the planet," as they say in science-fiction movies: I am, of course, referring to the famously silly (but undeniably awesome) 1990s movie *Hackers*, where two characters have a TV show called *Hack the Planet*.

217 Futurist Jamais Cascio: Personal interview, July 9, 2012. You can read *Hacking the Earth: Understanding the Consequences of Geoengineering* online: http://openthefuture.com/2009/02/hacking_the_earth.html.

217 sulfur-laced aerosol exhaust emitted: There are a number of studies showing a connection between ship aerosols and changes in the albedo of clouds. For example, P. A. Durkee et al., "The Impact of Ship-Produced Aerosols on the Microstructure and Albedo of Warm Marine Stratocumulus Clouds: A Test of MAST Hypotheses 1i and 1ii," *Journal of the Atmospheric Sciences* 57 (February 12, 1999): 2554–69. Using this as a form of geoengineering is discussed in part in Y.-C. Chen et al., "Occurrence of Lower Cloud Albedo in Ship Tracks," *Atmospheric Chemistry and Physics Discussions* 12 (2012): 13553–80. It remains unclear whether these ship aerosols are actually having a substantial effect on clouds and weather. See K. Peters et al., "A Search for Large-scale Effects of Ship Emissions on Clouds and Radiation in Satellite Data," *Journal of Geophysical Research* 116 (2011): D24205.

218 One of its researchers is Simon Driscoll: Personal interview, May 11, 2012.

219 The Harvard physicist and public-policy professor David Keith: See David W. Keith, "Photophoretic Levitation of Engineered Aerosols for Geoengineering," *Proceedings of the National Academy of Sciences* 107 (September 7, 2010): 16428–31.

219 Driscoll's colleagues at Oxford believe: See, for example, an experiment proposed by the group Driscoll works with, called Stratospheric Particle Injection for Climate Engineering (SPICE), in which

researchers have suggested they would accomplish atmospheric injection via a balloon tethered to the ocean: http://www2.eng.cam.ac.uk/~hemh/SPICE/SPICE.htm.

219 Alan Robock has run a number of computer simulations: Alan Robock, "20 Reasons Why Geoengineering May Be a Bad Idea," *Bulletin of the Atomic Scientists* 62 (May/June 2008): 14–18.

220 During many of the experiments, however: This difficulty is explored in the Royal Society report *Geoengineering the Climate: Science, Governance and Uncertainty* (London: the Royal Society, 2009). More recent experiments with iron fertilization appear to work somewhat better than the ones described in the Royal Society report. See Victor Smetacek et al., "Deep Carbon Export from a Southern Ocean Iron-Fertilized Diatom Bloom," *Nature* 487 (July 2012): 313–19. Smetacek and his colleagues report that the algae they worked with fell over 1,000 meters into the deep ocean, often finding its way into sediment on the ocean floor.

221 Tim Kruger, who heads the Oxford Martin School's geoengineering efforts: Personal interview, May 10, 2012.

222 The Cambridge physicist David MacKay: See David MacKay, *Sustainable Energy–Without the Hot Air* (Cambridge: UIT Cambridge, Ltd., 2009).

CHAPTER TWENTY: NOT IN OUR PLANETARY BACKYARD

226 Torino scale, a kind of Richter scale: Devised by MIT astronomer Richard Binzel in the late 1990s, the Torino scale is used to estimate the likelihood that an object will hit the Earth, as well as how much damage it might do. You can see a diagram of the Torino scale here: http://impact.arc.nasa.gov/torino.cfm.

226 NASA launched an asteroid-spotting program called Spaceguard: Read more about Spaceguard at NASA: http://neo.jpl.nasa.gov/neo/report.html.

227 bring them close to our own orbit: Generally, an object is classed as an NEO if its orbit is between 0.983 and 1.3 AU from the Sun. One AU (astronomical unit) is the average distance from the Earth to the Sun, or 149,597,871 kilometers.

227 some near misses: For a complete list of every NEO that's flown by since 1900, you can search NASA's database here: http://neo.jpl.nasa.gov/cgi-bin/neo_ca.

227 asteroid hunter Amy Mainzer calls one of the most hopeful: Personal interview, July 9, 2012.

228 she and her colleagues estimate: A. Mainzer et al., "Characterizing Subpopulations Within the Near-Earth Objects with NEOWISE: Preliminary Results," *Astrophysical Journal* 752 (June 20, 2012): 110.

230 Run by aerospace engineer William Ailor: Personal interview, July 9, 2012.

230 a series of suggestions over the past 15 years for how we'd deal with

asteroid threats: These include, among others, recommendations urging governments to gather more data on the locations of asteroids, create a governmental organization for "planetary defense," fund tests to figure out how we could move an asteroid, and "include NEO impacts as possible disaster scenarios for disaster recovery and relief agencies."

231 Ailor's company, the Aerospace Corporation, did a study in 2004: You can read the entire study via NASA here: http://impact.arc.nasa.gov/news_detail.cfm?ID=139.

232 NASA's Deep Impact mission: Learn more about Deep Impact via NASA here: http://www.nasa.gov/mission_pages/deepimpact/main/index.html.

233 In an early paper about nuclear winter: Alan Robock, "Snow and Ice Feedbacks Prolong Effects of Nuclear Winter," *Nature* 310 (1984): 667–70.

234 Alex Weir, a software engineer based in Zimbabwe: Download the database here: http://www.cd3wd.com/.

234 CD3WD and similar projects can help us restart civilization: One of these is Marcin Jakubowski's Civilization Starter Kit project. It's a free online resource (you can print it out to prepare for an asteroid strike) that will eventually contain all the information required to make 50 crucial farming machines and sustain a small village. Find out more here: http://opensourceecology.org/gvcs.php. A slightly different example is the Svalbard Global Seed Vault, located beneath the mountains on Norway's remote Arctic island of Svalbard, where philanthropic and diplomatic groups have paid to build a massive underground structure for seed storage. It's designed to be a backup copy of the planet's ecosystems in the event of a dramatic crash in biological diversity, such as what you'd see after a global disaster like a PHO impact. See more at: http://www.regjeringen.no/en/dep/lmd/campain/svalbard-global-seed-vault.html?id=462220.

CHAPTER TWENTY-ONE: TAKE A RIDE ON THE SPACE ELEVATOR

236 NASA has offered prizes of up to $2 million: Find out more about NASA's Strong Tether Challenge, which offers up to $2 million to the team that creates an elevator cable strong enough to form the centerpiece of a space elevator, on the NASA website: http://www.nasa.gov/offices/oct/early_stage_innovation/centennial_challenges/tether/index.html.

Much like the X Prize, the Space Elevator Games are events where inventors compete for large cash prizes, in this case for viable models of space-elevator climbers and ribbon structures. ISEC, the International Space Elevator Consortium, is a group that holds annual conferences and prize giveaways for inventors and investors to explore novel materials and methods we could use to build a space elevator. This association unites groups from Europe, Japan, and America that are working on space-elevator engineering, and they also publish books and a journal devoted to designing a working space elevator.

236 a scientist named Bradley Edwards, who wrote a book about the feasibility: Though there are other models for space elevators, Edwards's design is the one that NASA and its affiliated scientists are currently pursuing. Indeed, spaceflight engineer Peter Swan has written extensively about the process of building space elevators, and suggests that the final designs may change a great deal by the time we're actually building the structure. See, for example, Peter Swan and Cathy Swan, *Space Elevator Systems Architecture* (Lulu .com, 2007).

237 reduction in water pollution from perchlorates: See the EPA's information about perchlorates: http://water.epa.gov/drink/contaminants/unregulated/perchlorate.cfm/.

237 NASA reports that each Space Shuttle launch cost about $450 million: See NASA's Space Shuttle FAQ online: http://www.nasa.gov/centers/kennedy/about/information/shuttle_faq.html#10.

238 100,000 kilometers out into space: There is some disagreement over whether this length is necessary, or whether it could be shorter. Different plans call for different lengths.

239 In 2009, NASA awarded $900,000 to LaserMotive: Clara Moskowitz, "Space Elevator Team Wins $900,000 from NASA," MSNBC.com (January 7, 2009).

241 "I like to compare [carbon nanotube development]": Personal interview, August 12, 2011.

241 Engineer Keith Lofstrom suggested: Personal interview, August 12, 2011: You can see his plans for the maglev platform on launchloop .com.

241 Vasilii Artyukhov argued that we might not want to use carbon nanotubes: "Making and Breaking Graphitic Nanocarbon: Insights from Computer Simulations," Space Elevator Conference presentation, Microsoft Campus, Richmond, WA (August 12, 2011).

242 "The bottom line is that you need to find": Personal interview, June 26, 2012.

243 Alex Hall, the senior director of the Google Lunar X Prize: "When Can I Buy My Ticket to the Moon?" Panel discussion at SETICon 11, Santa Clara, CA (June 23, 2012).

244 Bob Richards, a cofounder of Moon Express: Ibid.

CHAPTER TWENTY-TWO: YOUR BODY IS OPTIONAL

246 I visited the UC Berkeley synthetic biologist Chris Anderson: Personal interview, June 6, 2012.

248 "During a long-duration manned space flight": Personal correspondence, June 25, 2012.

249 Still, her research and that of other geneticists working with NASA:

Personal interview with Sylvain Costes, June 15, 2012. Costes works at the Berkeley Lab, researching how radiation damages DNA. His work is funded in part by NASA, and he hopes that at some point we can pinpoint regions on the genome responsible for making some people's DNA more robust against cosmic radiation damage.

249 "I'm of the mind that we're going to fuck everything up": Personal interview, May 7, 2012.

249 "pollution sensing lung tumor": You can see this design, along with several more from the Synthetic Kingdom series, on Ginsberg's website here: http://www.daisyginsberg.com/projects/synthetickingdom.html.

250 The British author Paul McAuley has suggested in recent novels: Personal correspondence, June 12, 2012. See also *The Quiet War* (Amherst, NY: Pyr Books, 2009).

251 Kim Stanley Robinson, another science-fiction author: Personal interview, June 18, 2012. Robinson's most recent novel about how synbio modifications will be part of space colonization is called *2312* (New York: Orbit Books, 2012).

252 Nick Bostrom heads the institute, where he's written widely cited articles: Personal interview, May 8, 2012. See also Bostrom's considerable body of work on this subject, starting with "When Machines Outsmart Humans," *Futures* 35 (2000): 759–64. You can read the full text of this essay, along with many others, on Bostrom's personal website: http://www.nickbostrom.com. I'd also recommend an essay collection Bostrom coedited with Milan Ćirković called *Global Catastrophic Risks* (Oxford: Oxford University Press, 2008). This is a book produced by the Institute for the Future of Humanity, and it introduces many of the key concerns the institute addresses, including the intelligence explosion.

254 "having a biological body in space is stupid": Personal interview, May 8, 2012.

255 Many evolutionary biologists believe that humans are still evolving: Many recent studies deal with how humans are still under selection. For example: Alexandre Courtiol et al., "Natural and Sexual Selection in a Monogamous Historical Human Population," *Proceedings of the National Academy of Sciences* 109 (March 28, 2012): 8044–49. Courtiol and his colleagues argue that a thorough examination of the lineages of a Finnish village reveals natural and sexual selection at work, producing people who meet definitions of fitness involving better resistance to disease. Other researchers look at the human genome, and have discovered that some genes are undergoing fairly rapid transformation. Bruce Lahn and his colleagues describe how two genes that regulate gene size appear to be rapidly evolving in humans: P. D. Evans, S. L. Gilbert, N. Mekel-Bobrov, E. J. Vallender, J. R. Anderson, et al., "Microcephalin, a Gene Regulating Brain Size, Continues to Evolve Adaptively in Humans," *Science* 309 (2005): 1717. John Hawks has also written about this in a paper with his colleagues: "Recent Acceleration of Human Adaptive Evolution,"

Proceedings of the National Academy of Sciences 104 (December 26, 2007): 20753–58.

255 I spoke to Oana Marcu, a SETI Institute biologist: Personal interview, June 23, 2012.

CHAPTER TWENTY-THREE: ON TITAN'S BEACH

259 "Our kids are the last generation": Personal interview, June 26, 2012.

259 Armin Kleinboehl is far more conservative in his estimates: I spoke with Kleinboehl on June 10, 2012, during the Jet Propulsion Laboratory's annual open house, a fantastic event where scientists meet members of the general public, give them tours of the facilities, and explain what people at the lab are studying. Find out more about the Mars Reconnaissance Orbiter here: http://science.jpl.nasa.gov/projects/MRO/.

259 Futurists like Ray Kurzweil: See, for example, Kurzweil's book *The Singularity Is Near: When Humans Transcend Biology* (New York: Penguin Books, 2006). Other futurists who suggest the future is speeding up include Nick Bostrom, whose work I discuss in chapter 22, and Bill Joy in his famous essay "Why the Future Doesn't Need Us," *Wired* 8.04 (April 2000). Among futurists, this idea is sometimes referred to as "Moore's law." The sobriquet was originally intended to describe how computer chips improve exponentially over time. Now it's used to describe any exponential growth in scientific knowledge over time.

260 a project run by the doctor and former astronaut Mae Jemison: Personal interview, June 23, 2012.

261 planetary scientist Nathalie Cabrol: Personal interview, June 23, 2012. For more about Cabrol's work in the high lakes, see N. A. Cabrol et al., "The High-Lakes Project," *Journal of Geophysical Research: Biogeosciences* 114 (2009): G00D06. She also has an incredible field log of some of her work there, which you can read here: http://www.highlakes.seti.org/.

262 led the celebrated science historian Richard Rhodes to speculate: He made this speculation on the panel "All Aboard the 100 Year Starship" at SETICon II (June 23, 2012). He was specifically referring to Jemison's work, but I think it's fair to say that Cabrol's is relevant here too.

ILLUSTRATION CREDITS

14 Illustration by Stephanie A. Fox
25 Illustration by Stephanie A. Fox
36 Illustration by Stephanie A. Fox
39 Illustration by John Sibbick
41 Peter Roopnarine, Jonathan Mitchell, and Kenneth Angielczyk
66 Illustration by Stephanie A. Fox
92 © The British Library Board. Royal MS 18.E.i-ii f. 175 (date: 1385–1400).
130 David Kilper for Washington University
136 ANIMALS ANIMALS © Bob Cranston
161 Illustration by Stephanie A. Fox
170 Courtesy O. H. Hinsdale Wave Research Laboratory, Oregon State University
172 Shakhzod Takhirov, Site Operations Manager of nees@berkeley, University of California at Berkeley
184 From *On the Mode of Communication of Cholera*, by John Snow, published by C. F. Cheffins, Lith., Southampton Buildings, London, England, 1854.
195 Tim Barker/Lonely Planet Images/Getty Images
203 (above) Glenn Beanland/Lonely Planet Images/Getty Images
203 (below) Photo by Robinson Esparza
223 © Guardian News & Media Ltd., 2011
231 Ron Miller
238 NASA Artwork by Pat Rawlings/Eagle Applied Sciences
258 *Cassini* Radar Mapper, JPL, ESA, NASA

INDEX

Page numbers in *italics* refer to illustrations.

ABOUT THE AUTHOR

ANNALEE NEWITZ is the founding editor of the science Web site io9.com and a journalist with a decade's experience writing about science, culture, and the future for such publications as *Wired, Popular Science,* and *The Washington Post*. She is the editor of the anthology *She's Such a Geek!: Women Write About Science, Technology, and Other Nerdy Stuff* and was a Knight Science Journalism Fellow at MIT. She lives in San Francisco.